An Atlas for Staging Mammalian and Chick Embryos

Authors

H. Butler, Ph.D.
Professor of Anatomy
Department of Anatomy
College of Medicine
University of Saskatchewan
Saskatoon, Canada

B. H. J. Juurlink, Ph.D.
Associate Professor
Department of Anatomy
College of Medicine
University of Saskatchewan
Saskatoon, Canada

CRC Press, Inc.
Boca Raton, Florida

Library of Congress Cataloging-in-Publication Data

Butler, H. (Harry)
    An atlas for staging mammalian and chick embryos.

    Bibliography: p.
    Includes index.
    1. Embryology--Mammals--Atlases. 2. Chick embryo--
Atlases. I. Juurlink, B. H. J., 1947—   . II. Title.
[DNLM: 1. Chick Embryo--growth & development--atlases.
2. Embryology--atlases. QS 617 B985a]
QL959.B94   1987   598'.617   86-9552
ISBN-0-8493-6629-1

Direct all inquiries to CRC Press, Inc., 2000 Corporate Blvd., N.W., Boca Raton, Florida, 33431.

© 1987 by CRC Press, Inc.

International Standard Book Number 0-8493-6629-1
Library of Congress Card Number 86-9552
Printed in the United States

PREFACE

During investigations of development it is frequently necessary to arrange a series of embryos in their correct order of development or to compare embryos of equivalent stages in different species. This can only be done by assigning each embryo to a specifically defined stage of development. Much data on various mammals which vary greatly in quantity and quality for different species are widely scattered in the literature. For the convenience of research workers, particularly those with limited knowledge of embryology, we have put together a series of tables to enable rapid identification of specific stages of embryonic development in the more commonly used mammals. Because of its frequent usage the chick embryo is included in this atlas.

THE AUTHORS

Harry Butler, M.D., is Professor Emeritus of Anatomy in the University of Saskatchewan, Saskatoon, Saskatchewan, Canada.

Dr. Butler graduated in 1938 from the University of Cambridge with a B.A. degree in natural sciences. He subsequently obtained the degrees of M.B., B.Chir. in 1942, M.A. in 1946, M.D. in 1950, and Ph.D. in 1976 from the University of Cambridge. After wartime service as a medical officer in the Royal Navy, he returned to the University of Cambridge in 1946 and began his career as an anatomist. In 1951 he became a Reader in Anatomy in the University of London. From 1955 to 1964 he was Professor and Head of the Department of Anatomy in the University of Khartoum, Sudan. From 1960 to 1963 he was the Dean, Faculty of Medicine. In 1964 he joined the University of Saskatchewan and retired as Professor Emeritus in 1984.

His early work concerned the venous system, with particular reference to its development. It was in the Sudan that he became interested in the reproduction and development of the lesser bushbaby and in 1983 published the definitive account of its embryology.

He has been a member of the Antatomical Society of Great Britian and Ireland, the British Medical Association, the Philosophical Society of the Sudan, the Canadian Association of Anatomists, the Canadian Federation of Biological Sciences, the International Primatological Society and the American Society of Primatologists.

Bernhard H. J. Juurlink is an Associate Professor in the Department of Anatomy, University of Saskatchewan, Saskatoon, Saskatchewan, Canada.

Dr. Juurlink graduated in 1969 from Acadia University, Wolfville, Nova Scotia, Canada with a B.Sc. degree in biology, and in 1971 he obtained a M.Sc. degree from the same department. In 1975, he obtained a Ph.D. from McMaster University, Hamilton, Ontario, Canada.

Dr. Juurlink is a member of the Canadian Association of Anatomists, the Canadian Association of Cell Biologists, the Society for Developmental Biology, the International Society for Developmental Neuroscience and the American Society of Zoologists.

ACKNOWLEDGMENTS

We wish to thank Irene Partridge and Brenda Peters for typing the manuscript and Osman Kademoglu for the photographic work.

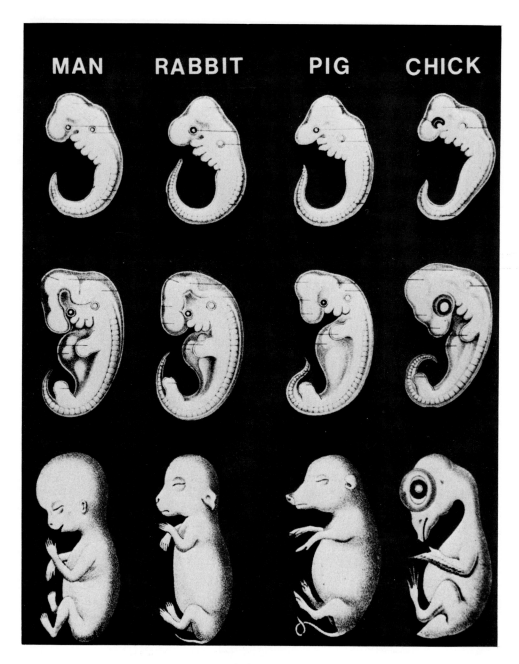

Embryos of different species pass through identical embryonic stages before acquiring their specific features (modified from Haeckel, E., *Anthropogenic ou Histoire de L'Evolution Humaine*, C. Reinwald et Cie, Paris, 1877.)

TABLE OF CONTENTS

INTRODUCTION

During experimental or observational investigations of mammalian development it is essential to be able to:

1. define a sequence of developmental stages for a given species which can be used to identify the stage of development of individual specimens.
2. utilize these developmental stages when comparing embryos of different species, since embryos of the same length or age are frequently at very different stages of development.

Embryos have been placed in a sequential order of development by measuring their length. Mall[1] arranged 266 human embryos, ranging from 2 to 25 mm in length into 14 stages according to their mean lengths. According to O'Rahilly[2] the most useful single measurement is " . . . the greatest length of the embryo as measured in a straight line (that is, caliper length) without any attempt to straighten up the natural curvature of the specimen" (Figure 1). Early embryos are straight but as development proceeds they develop an increasing ventral curvature forming a flattened spiral since the tail lies to one or other side of the head. Thus the greatest length, as defined by O'Rahilly,[2] becomes the crown rump length (C.R. length) which is shorter than the true greatest length (Figure 1). A slight change in the ventral curvature of the embryos will cause a considerable change in the C.R. length. The ventral curvature of the embryo is greatest during the period of organogenesis, a period of critical importance in teratological studies, and differences in ventral curvature are undoubtedly one reason for the variation in the C.R. length at any given stage in this period (Figure 2). During the fetal period the fetus begins to straighten up and the ventral curvature almost completely disappears (Figure 1). The C.R. length is now referred to as the sitting height and it becomes a more reliable indication of age. It has become customary to measure the length of embryos after fixation since human embryos obtained in the operating room or embryos of wild animals were fixed immediately before being transported to the appropriate laboratory for examination. For purposes of standardization, Streeter[3] recommended measurements to be made after 2 weeks in 10% formalin. It should be noted that the degree of shrinkage or swelling varies considerably from one fixative to another. The length of embryos at equivalent stages of development shows considerable species variation; e.g., the embryos of man, lesser galago, and mouse at the Carnegie stage 22 of development have a C.R. length of 25.0, 12.0, and 11 to 12.0 mm, respectively. Thus it is clear that length alone is a misleading and unreliable way of classifying embryos, and such expressions as " . . . at the 18.0 mm stage" should never be used. However, the length of an embryo should be accurately recorded after fixation, since it acts as a pointer to its appropriate developmental stage.

Theoretically, the age of an embryo dated from a known time of ovulation or an estimated time of fertilization should be the most reliable method of staging embryos but in practice there are many pitfalls. Firstly, embryos develop at variable rates after fertilization, and the stated days of gestation must be regarded as modal and not inclusive ranges.[4] In humans the timing of gestation is further complicated by the unreliability of data on the menstrual cycle and the time of coition and particularly by the lack of obvious and overt signs indicating ovulation. O'Rahilly[2] has drawn up what may be regarded as the most accurate table of the age of human embryos. The aging of embryos of laboratory animals is more accurate since the female usually shows clear indications of impending ovulation. Animals may be, therefore, paired at the appropriate time for a specific length of time or until copulation is observed. Then a vaginal

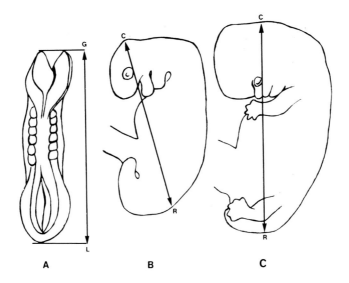

FIGURE 1.   Sketches illustrating methods of measuring the length of embryos. A = greatest length. B and C = crown rump length. Note how the increasing ventral curvature of the embryo decreases the greatest length as defined by O'Rahilly.[2]

smear can be examined for spermatozoa or in some species, e.g., rat and mouse, a vaginal plug is observed. However, as Juurlink and Fedoroff showed,[5] there may be considerable variation in the stage of development reached by embryos at any particular day of gestation (Figure 3). This variation was found to be an intralitter as well as interlitter phenomenon. The reasons for these differences are not clear, but differences in time of implantation and individual rates of development are probably involved. However, it is clear that gestational age is not an entirely reliable indication of the stage of development reached by an individual embryo. Another source of error is the manner in which the time of mating is used to estimate the commencement of embryonic development. Some authors call the day on which sperm or vaginal plug are found day zero of gestation, others, day one of gestation. In this work the day on which sperm or a vaginal plug is found will be regarded as day one of gestation.

In mammals it is customary to divide gestation into three main periods:

1.   The period culminating in implantation of the blastocyst and during which the fetal membranes are established and the germ layers are laid down in the embryonic disc.
2.   The embryonic period during which there is rapid growth and differentiation resulting in the laying down of all the main systems and organs of the body, and the major features of external body form are established.
3.   The fetal period which terminates at birth. This is a period of rapid increase in size associated with quite slow changes in body form. It is also a period of histological differentiation and the onset of function.

It is to be remembered that development is a continuous process and that these divisions of gestation are arbitrary.

Embryonic Staging

It has long been realized that there is a need for defining standardized stages in embryonic development of various organisms for the purpose of accurate description

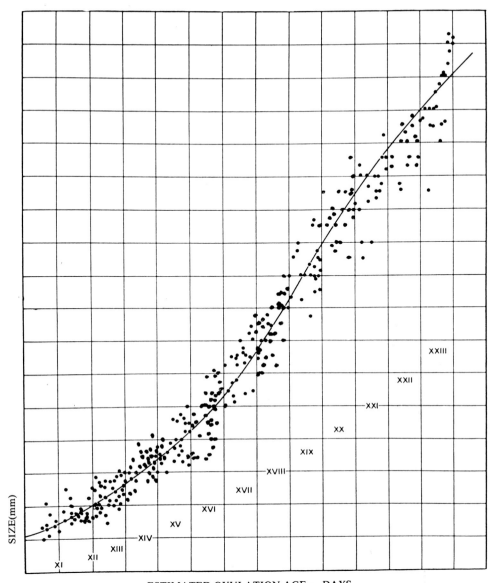

SIZE(mm)

ESTIMATED OVULATION AGE — DAYS

FIGURE 2.   Graphic plot of specimens in the Carnegie Embryological
Collection which have been surveyed and assigned to horizons XI to
XXIII, i.e., Carnegie Stages 11 to 23 under the new terminology. Note
the considerable spread in length at all stages but particularly in the older
stages. (From Streeter, G. L., *Contrib. Embryol. Carneg. Inst.,* 34, 165,
1951. Carnegie Institution of Washington, Department of Embryology,
Davis Division. With permission.)

of normal development and to enable accurate comparisons to be made between development in different species. Such stages are arbitrarily defined segments of the continuous process of development. O'Rahilly[2] defines embryonic stages thus: "Stages are
based on the apparent morphological state of development, and hence are not directly
dependent on either chronological age or on size. Furthermore, comparison is made of
a number of features of each specimen, so that individual differences are rendered less
significant and a certain latitude of variation is taken into account."

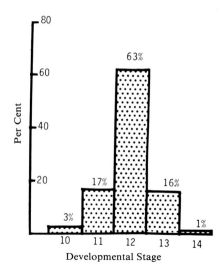

FIGURE 3. Proportion of mouse embryos at various developmental stages on day 8 of gestation. No. of litters = 13, no. of embryos = 135. Note, in this graph the morning of the vaginal plug is considered day 0 of gestation. (Modified from Juurlink, B. H. J. and Fedoroff, S., *In Vitro*, 15, 86, 1980.)

Although embryonic staging was introduced towards the end of the nineteenth century, there is no doubt that G. L. Streeter's "Developmental Horizons in Human Embryos"[6-10] forms the basic guideline for staging mammalian embryos. He divided human embryonic development into 23 stages or developmental horizons. He only considered horizons XI to XXIII since there was, at that time, insufficient material available for the first ten horizons. The onset of marrow formation in the humerus is rapid and easily observed and was "arbitrarily adopted as the conclusion of the embryonic and the beginning of the fetal period of prenatal life: It occurs in specimens about 30 mm in length".[9] The transition from embryo to fetus is also defined as the time of closure of the secondary palate.[4] Horizon X was described by Heuser and Corner in 1957.[11] The earlier horizons were described by O'Rahilly[2] who introduced a new nomenclature and re-assessed the age of the embryos in the light of more recent investigations. He replaced the term "horizon" by "stage" because the latter is the simple term employed for all other vertebrate embryos and had been applied to human embryos as early as 1914 by Mall.[1] He concludes: "The term is simpler, clearer, of widespread usage, and can be employed as a verb (to stage an embryo) as well as a participial adjective (a staging system)". Furthermore, it should be pointed out that such expressions as "at the 3 mm stage", should be replaced by "at 3 mm". In other words, the length of an embryo is a single criterion that is not in itself sufficient to establish a "stage". Other changes include replacement of Roman by Arabic numbers and elimination of the scientifically meaningless term "ovum" which was so frequently applied to pre-implantation stages. Thus, now the embryonic development of the human is divided into Carnegie stages 1 to 23.

On the basis of somite number and internal characters, it is possible to arrange chick embryos into Carnegie stages 9 to 23. These stages can be recognized on the basis of their length, external characters, and incubation age. Thus it is now possible to compare mammalian and chick embryos at equivalent stages of development.

This atlas of mammalian and chick embryos provides a means of staging them during the period of organogenesis, i.e., from the appearance of the somites to the tran-

Table 1
DIAGNOSTIC FEATURES OF MAMMALIAN EMBRYOS — FROM CARNEGIE STAGES 9 TO 23

Carnegie stage	Features
9	1 to 3 pairs of somites; open neural plate
10	4 to 12 pairs of somites; neural folds begin to fuse; 2 branchial bars; otic placode invagination
11	13 to 20 pairs of somites; cranial neuropore closes
12	21 to 29 pairs of somites; caudal neuropore closes; C-shaped embryo; 3 branchial bars; beginning cranial limb buds
13	4 limb buds present; maxillary and mandibular processes; closed otocyst; flat olfactory placode; tail bud
14	Cervical flexure; Nackengrube; open lens pit; definitive tail
15	Forelimb footplate; nasal pit; closed lens vesicle; early auricular hillocks; umbilical hernia
16	Retinal pigment; hindlimb footplate; nasal pits face ventrally
17	Finger rays; 6 distinct auricular hillocks; nasolacrimal groove
18	Notched handplate; elbow; toe rays; beginning eyelids; vibrissae; mammary buds
19	Early notched footplate; limbs extend nearly straight forward; head begins to be raised beyond a right angle; continued fusion of auricular hillocks
20	Arms increased in length and bent at elbow; forelimb digits separating; hindlimb digits visible; eyebrow follicles
21	Forelimb digits longer and have touch pads; feet approach each other; helix and antihelix
22	Thickened eyelids encroaching on eyeballs; tragus and antitragus
23	Fusion of secondary palate; limbs longer and more developed; all digits completely separated; eyelids cover most of eye; definitive shape of auricle; hair follicles on body

sition stage from embryo to fetus. The standard used is Carnegie stages 9 to 23 of human embryos, and all other embryos are, wherever possible, compared to these. Ideally, it is necessary to analyze the following data for each embryo under consideration:

1. Greatest length as defined by O'Rahilly[2]
2. Known or estimated postovulatory age
3. Description of external features
4. Serial sections of the embryo are examined to establish the stage of development of the various diagnostic organs.

The diagnostic organs vary from stage to stage of development and are fully described for human embryos by Streeter[6-10] and O'Rahilly.[2]

This degree of examination is, however, only required when staging embryos of any particular species for the first time. For the purpose of everyday laboratory identification of embryonic stages, only greatest length, known or estimated age, and external features will be used. Tables 1 and 2 list the diagnostic features of mammalian and chick embryos.

Table 2

DIAGNOSTIC FEATURES OF CHICK EMBRYOS — FROM CARNEGIE STAGES 9 TO 23

Carnegie stage	Features
9	1 to 3 pairs of somites; open neural plate
10	4 to 12 pairs of somites; neural folds begin to fuse; otic placode; closure of caudal neuropore
11	13 to 20 pairs of somites; 2 branchial bars; otic pits; cranial neuropore closed
12	21 to 29 pairs of somites; closure of caudal neuropore; 3 branchial bars; olfactory placode; distinct cranial limb buds
13	4 limb buds present; tail bud; otocyst closed
14	35 to 39 pairs of somites; lens vesicle closes; maxillary process; deep olfactory pit
15	Increased length of limb buds; very prominent maxillary process; retinal pigment appearing
16	Toe plate on leg bud; 6 distinct auricular hillocks; cervical sinus indistinct; umbilical hernia
17	Leg bud longer than wing bud; elbow and knee present; toe rays beginning in foot plate; most ventral auricular hillock forms the "collar"
18	Distinct toe rays; the "collar" is more distinct, but remaining auricular hillocks are beginning to fuse; distinct beak by end of stage
19	Leg still markedly longer than wing; both have distinct digital rays; beak more prominent and egg tooth appearing; distinct eyelids; increased length of neck between "collar" and mandible; other auricular hillocks flattened and fusing; never more than 2 scleral papillae; feather germs
20	All digits delineated and considerably lengthened; anterior tip of mandible has reached beak; "collar" disappearing; up to 8 scleral papillae; more feather germs
21	Toes separating and first digit of wing distinct; webs between digits; 9 to 14 scleral papillae and circle complete; nictitating membrane well developed; more feather germs
22	Fingers and toes more separated; webs disappearing; all feather germs more conspicuous; eyelids forming
23	Secondary palate fused; eyelids beginning to close; toes completely separated and claws appearing

Table 3
CARNEGIE STAGES 1 TO 8 OF HUMAN EMBRYOS

Carnegie stage	Age (days)	Features
1	1	Fertilization
2	2—3	From 2 to about 16 cells
3	4—5	Free blastocyst
4	5—6	Attaching blastocyst
5	7—12	Implanted though previllous
6	13—15	Chorionic villi; primitive streak may appear
7	15—17	Notochordal process
8	17—19	Primitive pit; notochordal and neurenteric canals

Based on O'Rahilly, R., *Developmental stages in human embryos, including a survey of the Carnegie collections. Part A. Embryos of the first three weeks.* (Stages 1 to 9), Carnegie Institute of Washington, Washington, D.C., 1973.

I. MAN

Carnegie Stages of Human Embryos [*Homo sapiens*]
The 23 Carnegie stages of human embryos will be used as the reference base in this atlas and will be presented in two ways:

1. The presomite stages (Carnegie stages 1 to 8) are listed (Table 3) but not illustrated because: (a) teratological research is rarely, if every carried out during this period; (b) staging requires microscopic examination; and (c) staging is largely dependent upon details of implantation which are not comparable with those of other species.
2. Stages 9 to 23 cover the all-important period of embryogenesis. These stages will be illustrated and a brief account of the internal features diagnostic of each stage will be given. The criteria used to identify a particular stage are: (a) greatest length, (b) known or estimated age, and (c) external features. In general, the greatest length and external features will be the most reliable criteria.

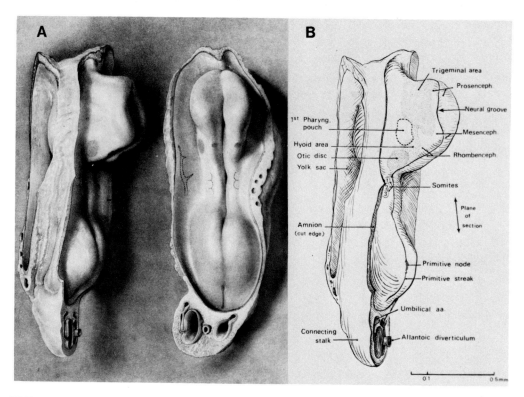

FIGURE 4.   (A) Left lateral and dorsal views of a reconstruction of Carnegie embryo No. 1878. (B) Left lateral view of the same embryo showing ectodermal areas of head region (stippled) and the mesencephalic flexure. (From O'Rahilly, R., *Developmental stages in human embryos, including a survey of the Carnegie collections. Part A. Embryos of the first three weeks (Stages 1 to 9),* Carnegie Institute of Washington, D.C., 1973. With permission.)

Man. Carnegie Stage 9

Embryos of stage 9 vary from approximately 1.5 to 2.5 mm in length and have a postovulatory age of 20 days. The characteristic feature of this stage is the appearance of from one to three pairs of somites. As seen from the dorsal aspect, the embryo is frequently described as having the shape of the sole of a shoe. Many embryos have a dorsal concavity or "lordosis". Very abrupt kinks are probably abnormal. The neural groove is now quite deep but open along its complete length. The cranial half of the neural groove represents the future brain. The elevated cranial ends of the neural folds are separated by a terminal notch which leads to the buccopharyngeal membrane. In more advanced specimens the cranial (or mesencephalic) flexure appears at the midbrain. The otic disc (or plate) appears during this phase, but the optic primordia are not visible. A primitive groove may be found but a distinct node is not always recognizable. The yolk sac has numerous blood islands and a vitelline plexus is visible.

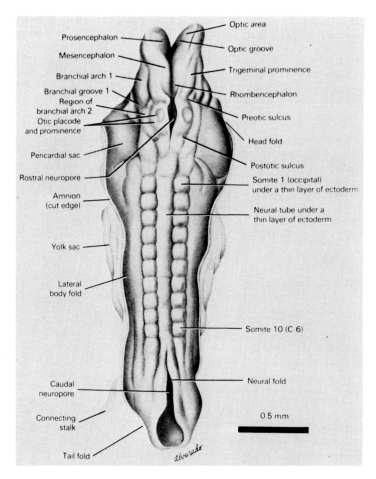

FIGURE 5. Dorsal view of a 10 somite embryo. (From Gasser, R. F., Atlas of Human Embryos, Harper and Row, Maryland, 1975. With permission.)

Man. Carnegie Stage 10

External Features

Embryos of stage 10 vary from 2.0 to 3.5 mm in length and have a postovulatory age of 22 days. They have from 4 to 12 pairs of somites. This stage is marked by the commencement of the fusion of the neural folds. It begins in embryos with 6 to 7 pairs of somites and by the end of the period extends from the otic region to the level of the latest formed somite. The otic disc is beginning to invaginate. There is considerable elongation of the embryo and expansion of the yolk sac. The mandibular and hyoid bars begin to be visible externally. The head and tail folds appear. Towards the end of this period the increasing size of the heart makes the pericardial region a prominent feature of the external form.

Internal Features

The heart endothelium is enclosed in a jelly-like envelope which in turn is enclosed in a layer of contractile tissue, the primordium of the myocardium. All the blood vessels consist only of simple endothelium and contain very few blood cells. The optic evaginations and the primordia of the corpus striatum, thalamus, and tegmentum are present. The parts of the neural crest are emerging. Pronephric rudiments appear at 8 somites, and the first definitive mesonephric vesicle at 10 somites. There is a solid mesonephric duct which extends for a varying distance caudal to somite ten.

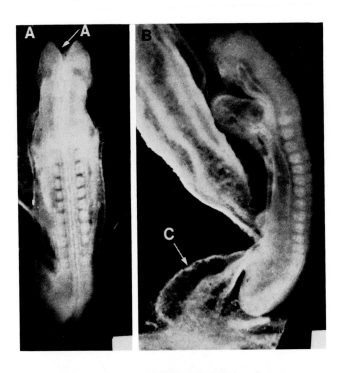

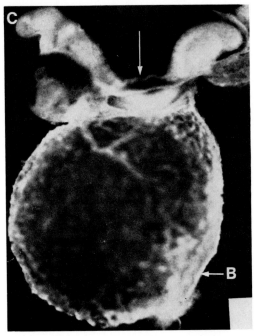

FIGURE 6.   (A) Dorsal view of a 13-somite embryo (Carnegie embryo No. 6344). (B) Left lateral view of the same embryo. (C) Left lateral view of a 19-somite embryo (Carnegie embryo No. 6050) to show the dorsal concavity or "lordosis" so frequently seen at this stage (arrow). A = cranial neuro pore; B = yolk sac; C = body stalk. (From Streeter, G. L., *Contrib. Embryol. Carneg. Inst.*, 30, 211, 1942. Carnegie Institution of Washington, Department of Embryology, Davis Division. With permission.)

Man. Carnegie Stage 11

External Features

Embryos of stage 11 vary from 2.0 to 4.5 mm in length and have a postovulatory age of 24 days. They have from 13 to 20 pairs of somites. This is the period of formation and closure of the cranial neuropore. In young specimens fusion of the neural plates has reached the level of the colliculi, while in the oldest specimens the cranial neuropore is closed or just closing. The mandibular and hyoid bars are well developed. The otic invagination can be recognized in most specimens as a slight depression, and in transparent specimens it can be clearly seen owing to the refraction of its thick margins. Many specimens of this stage have a marked "lordosis" or ventral convexity in the mid-body region and, consequently, accurate measurement is difficult to make.

Internal Features

This is the period of the proliferating mesoblastic cells that Streeter[6] called the coelomic tract. The blood vessels are still simple endothelial tubes, and only in the heart region are the auxiliary tissues differentiated. Foci of angiogenetic cells are widespread in the coelomic tract and on the surface of the central nervous system. The trigeminal and acousticofacial nerves form conspicuous proturberances. The primordia of the corpus striatum, thalamus, and tegmentum are present. Opposite the cranial somites the spinal neural crest has migrated ventralwards, but at the more caudal levels the neural crest cells are still incorporated in the roof of the neural tube.

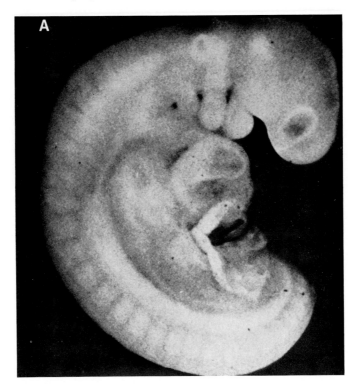

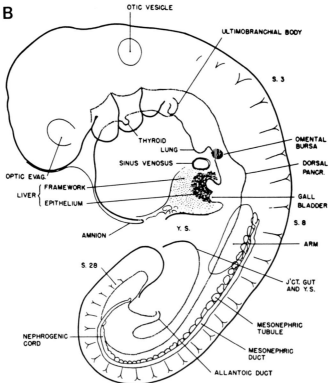

FIGURE 7.   (A) Right lateral view of a 29-somite embryo (Carnegie embryo No. 1062). (B) Profile reconstruction of the gut tract of a 28-somite embryo (Carnegie embryo No. 5923). (From Streeter, G. L., *Contrib. Embryol. Carneg. Inst.*, 30, 211, 1942. Carnegie Institution of Washington, Department of Embryology, Davis Division. With permission.)

Man. Carnegie Stage 12

External Features

Embryos of stage 12 vary from 3.0 to 5.0 mm in length and have a postovulatory age of 26 to 30 days. They have 21 to 29 pairs of somites. There is now a marked change in the shape of the embryo. Increase in the bulk of the spinal cord, somites, and mesonephros fill out the back of the embryo which now becomes C-shaped. The dorsal kink of previous stages is now no longer seen. Three branchial bars are now clearly visible and are beginning to be divided into dorsal and ventral parts. Caudal to the third bar is a depression, the cervical sinus. Closure of the caudal neuropore is complete in the oldest specimens. The otic pits are almost closed. A slight elevation in the older specimens indicates the beginnings of the cranial limb bud. The yolk stalk is appearing.

Internal Features

Key features of this stage are sharply outlined fields of epithelial proliferative activity marking the location of such primary organs as liver, thyroid, ultimobranchial bodies, stomach, liver, and dorsal pancreas. The liver is the most precocious of these organs and shows active angiogenesis. Rathke's pouch is present. Vascular specializations include: (1) changes at the venous end of the heart; (2) vascularization of the central nervous system; (3) establishment of the cardinal veins; (4) the hepatic plexus, and (5) formation of the vitelline arteries and veins. In the central nervous system the tegmentum is readily recognizable and there is thinning of the roof of the hindbrain. Neural crest development continues noticeably in the cervical region and from the wall of the otic vesicle. Mesonephric tubules extend from somite eight to somite 20, beyond which there is a continuous nephrogenic cord. Two or more tubules lie opposite each somite.

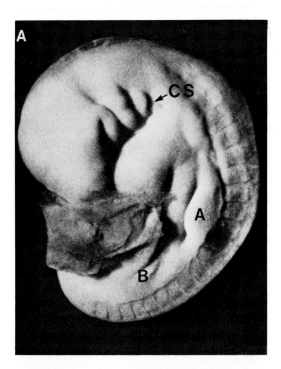

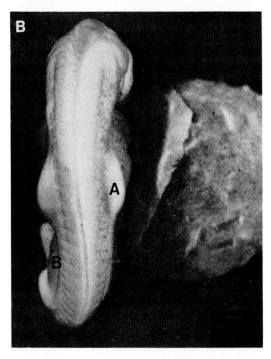

FIGURE 8.   (A) Left lateral view of 4.2-mm-long embryo (Carnegie embryo No. 7889). Its shortness is due to being coiled more closely than usual. CS = cervical sinus; A = cranial limb bud; B = caudal limb bud. (B) Dorsal view of 5.3-mm embryo (Carnegie embryo No. 8066). Note shape of cranial limb bud. (C) Ventral view of 5.2-mm embryo (Carnegie embryo No. 7433). YS = yolk sac; ST = yolk sac stalk. (From Streeter, G. L., *Contrib. Embryol. Carneg. Inst.*, 31, 27, 1945. Carnegie Institution of Washington, Department of Embryology, Davis Division. With permission.)

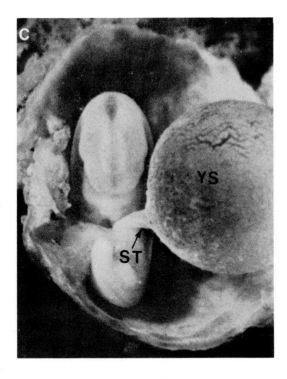

FIGURE 8C.

Man. Carnegie Stage 13

External Features

Embryos of stage 13 vary from 4.0 to 6.0 mm in length and have a postovulatory age of 28 to 32 days. The cranial limb buds form definite ridges and the caudal limb buds are beginning to appear. The central nervous system is still the main determinant of body form. There are three well-developed branchial bars, and the cervical sinus forms a triangular depression whose floor is clearly visible. The first branchial bar is dividing into maxillary and mandibular parts. The heart and liver are very prominent. There is a slender yolk stalk with markedly decreased transverse diameter. The otocyst is closed, and now the olfactory placode is visible. There is a distinct tail bud.

Internal Features

The various endodermal proliferations which began in the previous stage continue, and the dorsal pancreas is beginning to be constricted from the intestinal tube. The lung bud bifurcates into the primary bronchi, and the bilateral bronchial asymmetry is apparent. The vascular system continues to increase in size and complexity, and the circulation is well established. The gelatinous reticulum of the heart is now limited to three areas of narrowing which form valve-like junctions that prevent backflow at the sinoatrial foramen, the atrioventricular canal, and along the bulbus cordis and aortic trunk. Since the previous stages, the central nervous system has increased in size rather than in complexity. The cervical flexure is now present. The optic vesicle is in contact with the overlaying skin ectoderm, with little or no intervening mesenchyme. The lens placode is thickening but has not yet begun to invaginate. The endolymphatic duct is just beginning. The mesonephric duct opens into the cloaca.

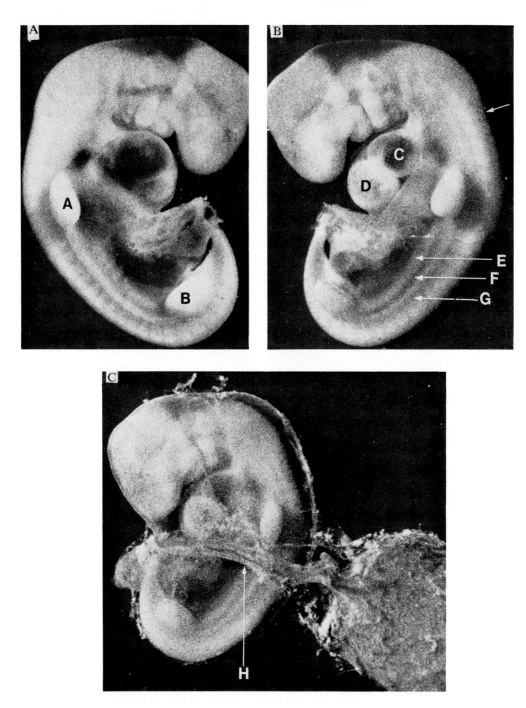

FIGURE 9.   (A,B,C) Three different views of a 7.3-mm embryo (Carnegie embryo no. 8141). A = cranial limb bud; B = caudal limb bud; C = atria; D = ventricles; E = unsegmented band of mesenchyme; F = posterior cardinal vein; G = mesonephros; H = yolk sac stalk. Arrow points to the Nackengrube. (From Streeter, G. L., *Contrib. Embryol. Carneg. Inst.*, 31, 27, 1945. Carnegie Institution of Washington, Department of Embryology, Davis Division. With permission.)

Man. Carnegie Stage 14

External Features

Embryos of stage 14 are 5.0 to 7.0 mm long and have a postovulatory age of 31 to 35 days. As before, the contour of the embryo is largely determined by the central nervous system. Most have a depression in the dorsal contour at the level of the 5th and 6th somites, the Nackengrube of His. It began in the oldest embryos of the preceding group but from now on is a characteristic feature of embryos up to 30.0 mm long, i.e., to the end of the embryonic period. The cranial limb buds form rounded, projecting appendages curving forwards and inwards. They taper towards their tips. The caudal limb buds are smaller and somewhat fin-shaped. An unsegmented strip of mesenchyme runs between the limb buds, which is thought to give rise to abdominal muscles. In front of it lies the slender mesonephros, and given the right lighting conditions, the posterior cardinal vein can be seen between them. Thin-walled atria are clearly distinguishable from the trabeculated ventricles. Mandibular and hyoid bars are conspicuous, but the third is beginning to disappear into the cervical sinus. The lens pit is still open, and the olfactory plate is usually visible but still flat. The otocyst is completely separated and has a well-defined endolymphatic duct.

Internal Features

The auditory tube and the middle ear cavity are appearing and the primordia of the thymus, lateral thyroid, and superior and inferior parathyroids are well developed. The median thyroid diverticulum is bilobed. Rathke's pouch is a prominent feature of the roof of the mouth. The trachea is separated from the esophagus. An extensive capillary network which surrounds the lung buds is fed by irregular bilateral channels from the plexiform sixth aortic arch arteries. A short pulmonary vein connects this plexus to the left atrium. The mesonephros is well on its way developmentally but apparently not yet functional. The best-developed mesonephric vesicles show the beginnings of the glomerulus, secretory segment, and collecting duct. Vasculogenesis is beginning in the mesonephros, but there are no feeder arteries from the dorsal aorta. The ureteric diverticulum and its covering metanephric cap is present. The spinal cord shows three distinct zones; the marginal non-nuclear zone, the expanding intermediate zone, and the ventricular zone. The wall of the brain has numerous expanded areas forming the primordia of basic functional centers which now includes the cerebellar plates. Cranial nerves III and IV are becoming visible. The lens vesicle shows varying degrees of indentation, and there may be slight variations between the right and left vesicles; but, so long as one of them is definitely open to the surface of the embryo, it is included in this age group.

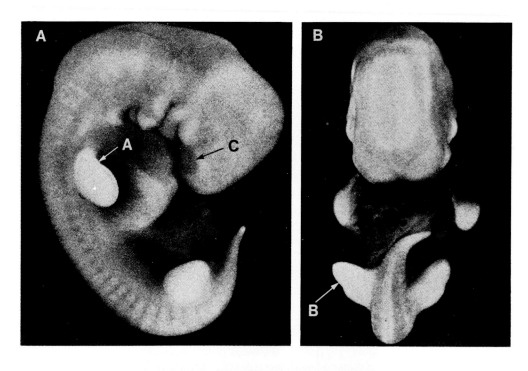

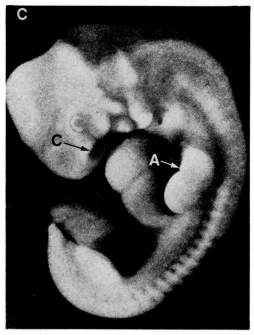

FIGURE 10.   (A) Right lateral view of a younger member of stage 15. Carnegie embryo No. 3441; 8.0 mm long. (B) Ventral view of an older member of stage 15. Carnegie embryo No. 3512; 8.5 mm long. (C) Left lateral view of previous embryo. A = groove separating hand plate from proximal part of arm; B = footplate; C = olfactory placode. (From Streeter, G. L., *Contrib. Embryol. Carneg. Inst.*, 32, 153, 1948. Carnegie Institution of Washington, Department of Embryology, Davis Division. With permission.)

Man. Carnegie Stage 15

External Features

Embryos of stage 15 are 7.0 to 9.0 mm long and have a postovulatory age of 35 to 38 days. Until this time, the central nervous system has been the principal determinant of the bodily contours, but now the heart, limb buds, and branchial bars begin to play a part. The trunk region is beginning to widen from side to side. Both lens vesicles are now completely closed. The olfactory placodes form large oval depressions facing more laterally. The hand and footplates are beginning to be delineated. Spreading mesenchymatous tissue is obscuring the somites and spinal ganglia. This begins in the occipital region and spreads caudalwards. The auricular hillocks appear on the mandibular and hyoid bars.

Internal Features

Stages 15 to 18 are primarily staged on features of the eye, ear, lung, and kidney (metanephros):

- Closure of the lens vesicle marks the beginning of stage 15.
- There is a distinct endolymphatic duct differentiated and set off from the remainder of the auditory sac.
- The sites of the secondary bronchi vary from slight swellings to short, stubby evaginations.
- The ureter has a distinct pelvis.

Additional features include a definite ileocaecal junction, a primary intestinal loop, and a ventral pancreas. The developing pineal can be located, but there is no sign of the neurohypophysis. The cerebral hemispheres have definite contours and an associated accentuation of the corpora striata. The rhombic grooves are still conspicuous.

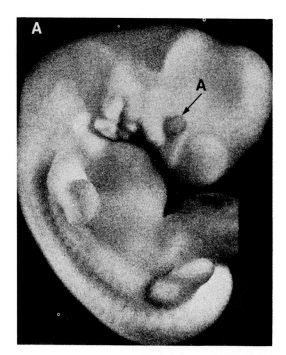
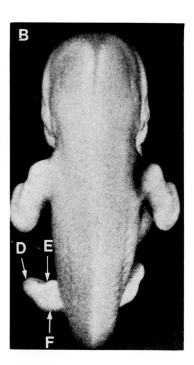
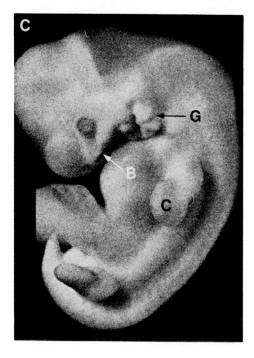

FIGURE 11.   Three views of an older embryo of stage 16. Carnegie embryo No. 8112; 10.9 mm long. A = retinal pigment; B = olfactory pit; C = handplate; D = footplate; E = lumbar growth center; F = sacral growth center; G = hyoid bar with auricular hillocks. (From Streeter, G. L., *Contrib. Embryol. Carneg. Inst.*, 32, 153, 1948. Carnegie Institution of Washington, Department of Embryology, Davis Division. With permission.)

Man. Carnegie Stage 16

External Features

Embryos of stage 16 are 8.0 to 11.0 mm long and have a postovulatory age of 37 to 42 days. The distinctive characters of this group are to be found in the nostril, eye, hyoid bar, and especially the limb buds. As development of the nostril progresses, the deepening olfactory pits face more and more ventralwards, and by the end of this stage their floor can no longer be seen in the lateral profile. The beginning of this stage is marked by the appearance of retinal pigment which can be distinctly seen in the older members of this group. The handplate is now differentiated into a central carpal part surrounded by a thick crescentic flange which will become the digital plate. The distal part of the leg bud forms the footplate, and the proximal part has lumbar and sacral growth centers roughly representing the thigh and leg. The ventral ends of the mandibular and hyoid bars are drawn inwards and disappear from view. The hyoid bar now becomes dominant because of the appearance of the auricular hillocks.

Internal Features

Changes in the four key characters are

- Definitive pigment granules are present in the outer coat of the retina.
- The relatively long endolymphatic duct is still a prominent feature, but the remainder of the vesicle is elongating to form the primordium of the cochlear pouch.
- The asymmetrical right and left stem bronchi end in bulbous terminal growth centers, and along the sides they give off elongating lateral branches.
- In older members the renal pelvis is bilobed.

The posterior lobe of the hypophysis is evaginating. The central part of the wide atrioventricular canal is showing fusion to produce the right and left canals.

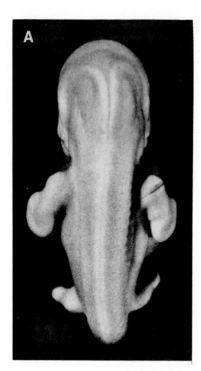
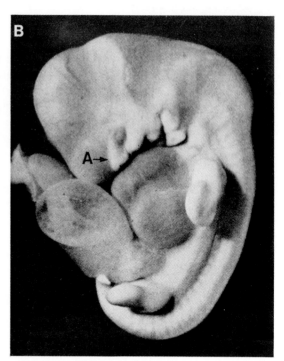

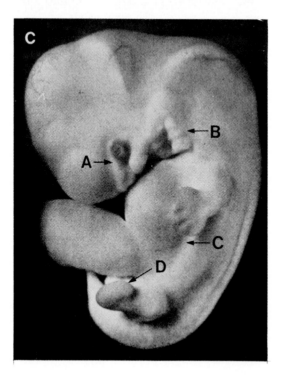

FIGURE 12.   (A and B) Two views of a younger embryo of stage 17. Carnegie embryo No. 8253; 11.2mm long. (C) Left lateral view of an older embryo of stage 17. Carnegie embryo No. 8101; 13.0 mm long. A = nasolacrimal groove; B = auricular hillock on mandibular and hyoid bars; C = finger rays; D = digital plate of leg bud. (From Streeter, G. L., *Contrib. Embryol. Carneg. Inst.*, 32, 153, 1948. Carnegie Institution of Washington, Department of Embryology, Davis Division. With permission.)

Man. Carnegie Stage 17

External Features

Embryos of stage 17 are 11.0 to 14.0 mm long and have a postovulatory age of 42 to 44 days. As a result of precocious growth of the brain the head forms about one half the profile area, a greater proportion than existed in any previous stage. The head and thoracic region are very wide, but the lumbosacral region is slender. The trunk has become straighter and begins to acquire a slight lumbar flexure. All specimens have a nasolacrimal groove (referred to as frontonasal or nasomaxillary groove by Streeter[8]). The auricular hillocks exhibit their characteristic form of 6 circumscribed superficial condensations, nos. 1 to 3 on the mandibular bar and 4 to 6 on the hyoid bar; the latter are the more prominent. The handplate has now acquired finger rays and in the older embryos begins to have a crenated edge due to projection of the tips of individual digits. The leg bud now has a rounded digital plate set off from the tarsal region and leg. Only the caudal somites are visible.

Internal Features

Changes in the four key features are

- Migration of retinal nuclei forming the first representatives of the inner nuclear layer restricted to approximately the position of the macula lutea. The lens cavity becomes a crescentic cleft.
- Three parts of the labyrinth wall are sinking preparatory to being absorbed to form the semicircular canals.
- The branches of the stem bronchi are acquiring branches of the second order.
- Well marked calyces are present in the renal pelvis.

The two atrioventricular canals are completely separated, and the pulmonary and aortic trunks are partially separated. The vermiform appendix is present, and the dorsal and ventral pancreases have fused but still have separate ducts. Bile ducts are beginning to invade the liver. The olfactory evagination is now definite, and the neural lobe of the hypophysis is present. The ostium of paramesonephric duct is defined.

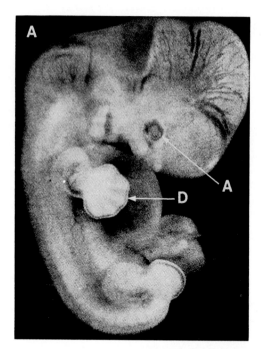
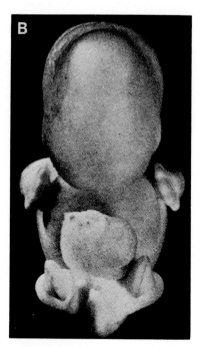
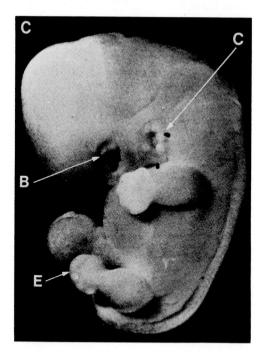

FIGURE 13.   (A) Right lateral view of a younger embryo of stage 18. Carnegie embryo No. 8097; 15.5 mm long. (B and C) Two views of an older embryo of stage 18. Carnegie embryo No. 7707; 14.5 mm long. A = optic cup; B = eyelid and conjunctival groove; C = merged auricular hillocks; D = digital notches; E = toe rays. (From Streeter, G. L., *Contrib. Embryol. Carneg. Inst.*, 32, 153, 1948. Carnegie Institution of Washington, Department of Embryology, Davis Division. With permission.)

Man. Carnegie Stage 18

External Features

Embryos of stage 18 are 13.0 to 17.0 mm long and have a postovulatory age of 44 to 48 days. [N.B. Cannot always distinguish old stage 17 from young stage 18 embryos on external form alone. Borderline specimens are graded on structure of selected internal characters.] Not only are stage 18 embryos larger, but they have advanced in the coalescence of their body regions which have now become more closely integrated in a common cuboidal bulk instead of consisting of individual and obviously separate parts. The optic cup has a quadrangular pigmented rim. Older embryos have eyelids and conjunctival grooves. The tip of the nose is visible. The auricular hillocks are merging to form a definitive auricle. Distinct finger rays and digital notches. Older embryos show bend of elbow. May show toe rays, but the footplate is not notched. Mammary buds are present.

Internal Features

Changes in the four key features are

- The heavily pigmented eye shows advanced migration forward of the inner nuclear layers of the retina.
- The membranous inner ear has one to three semicircular canals.
- The secondary branches of the bronchi are bifurcated.
- The renal calyces are branching and tubule formation is commencing, but there are no glomeruli.

The vomeronasal organ appears as a sharply invaginated groove. There is rapid growth of the paramesonephric duct. In some more advanced specimens, the sex of the gonad can be recognized. The aortic and pulmonary valves are clearly defined. The right and left ventricles are completely separated in the older specimens. The pineal has a rostral lobe and sometimes shows follicles.

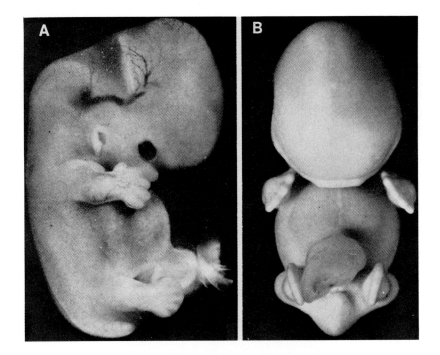

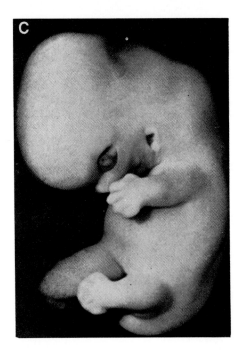

FIGURE 14.   Carnegie Stage 19 embryos. (A) No. 6824; 18.5 mm long. (B and C) Carnegie embryo No. 4501; 18.0 mm long. (From Streeter, G. L., *Contrib. Embryol. Carneg. Inst.*, 34, 165, 1951. Carnegie Institution of Washington, Department of Embryology, Davis Division. With permission.)

Man. Carnegie Stage 19

External Features

Embryos of stage 19 vary from 16.0 to 18.0 mm long and have a postovulatory age of 48 to 51 days. In the remaining stages, the embryo makes considerable progress in the development of its body form. In stage 19 the trunk region has begun to elongate and straighten out slightly, so that the head is no longer at a right angle to the back. The limbs extend nearly straight forward. In both arms and legs the various parts can be identified. The toe rays are more prominent, but there are no interdigital notches in the rim of the footplate. The auricular hillocks have become much less obvious as a result of continued fusion.

Internal Features

Streeter[10] devised a method of rating the development of selected organs on a system of point scores for stages 19 to 23. He chose key organs which are undergoing marked transformations readily recognizable under the microscope. The organs chosen were these: cornea, optic nerve, cochlea, hypophysis, vomeronasal organ, submandibular gland, kidney, cartilage, and bone. In the case of stage 19, they show the following features:

- A thin layer of loose mesenchyme, the primitive corneal body, crosses the midline of the eye.
- The small slender optic stalk has a lumen for practically all its length with few or no nerve fibers.
- The tip of the cochlea turns up but is short.
- The hypophysis has a thick stalk with a remnant of the lumen of Rathke's pouch; angiogenesis is beginning.
- The vomeronasal organ is essentially a groove or pit.
- The submandibular gland has a short, clublike duct entering the mesenchymal anlage of the gland.
- Metanephric vesicles are beginning to form.
- Differentiation of cartilage cells has progressed to phase three.

[Streeter[9] recognized five phases in cartilage development prior to ossification.]

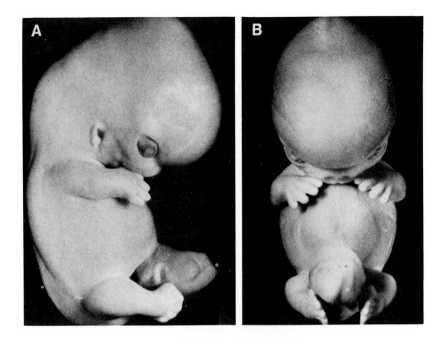

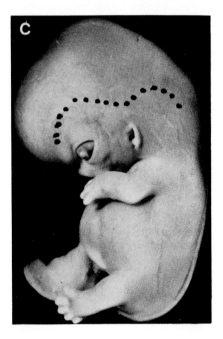

FIGURE 15.   Photographs of two embryos belonging to stage 20. (A) Carnegie embryo No. 7274; 18.5 mm long. (B and C) Carnegie embryo No. 8157; 20.8 mm long. Dotted line in C indicates advancing edge of vascular plexus. (From Streeter, G. L., *Contrib. Embryol. Carneg. Inst.*, 34, 165, 1951. Carnegie Institution of Washington, Department of Embryology, Davis Division. With permission.)

Man. Carnegie Stage 20

External Features

Embryos of stage 20 vary from 18.0 to 20.0 mm long and have a postovulatory age of 51 to 53 days. Thickening of the subcutaneous mesenchyme obscures the underlying structures, and parts of the brain which were evident in the previous stage are no longer visible. The arms increase in length and are increasingly bent at the elbow. The hands with their short stubby fingers are still far apart but are curving slightly over the heart region and approach the lateral margin of the nose. The toes are visible. All embryos have a delicate, fringe-like vascular plexus in the superficial tissue of the head. It centers over the eye and the ear, and the edge of the plexus is almost halfway between the eye-ear level and the vertex of the head.

Internal Features

The diagnostic internal features of Carnegie stage 20 are

- Corneal body is being differentiated from the remnant of crossing mesenchyme.
- A lumen present in practically the whole length of the optic stalk but few or no nerve fibers.
- Cochlea has a long tip and is in a transitional stage.
- Capillaries appearing in mesenchyme at rostral surface of anterior lobe of hypophysis; long, slender stalk.
- Broad opening at oral end of vomeronasal organ, but caudal end is a shallow blind sac.
- Long knobby duct well into the submaxillary gland.
- S-shaped lumen in renal vesicles.
- Clearing center in cartilage of bones.

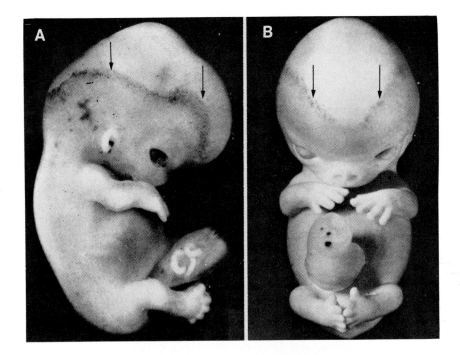

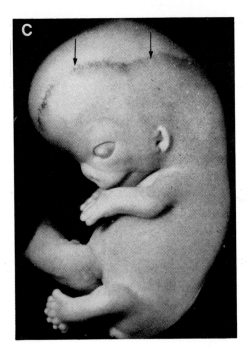

FIGURE 16.   Photographs of two embryos belonging to stage 21. (A) Carnegie embryo No. 4090. 22.2 mm long. (B and C) Carnegie embryo No. 7392, 22.7 mm long. Arrows indicate vascular plexus. (From Streeter, G. L., *Contrib. Embryol. Carneg. Inst.*, 34, 165, 1951. Carnegie Institution of Washington, Department of Embryology, Davis Division. With permission.)

Man. Carnegie Stage 21

External Features

Embryos of stage 21 vary from 22.0 to 24.0 mm in length and have a postovulatory age of 53 to 54 days. The superficial cranial vascular plexus is now somewhat more than half the distance from the eye-ear level to the vertex of the head. The fingers are longer and extend further beyond the ventral body wall than they did in the previous stage. The terminal phalanges are slightly swollen and show the beginnings of touch pads. The hands are slightly flexed at the wrists and nearly come together over the heart eminence. The feet are approaching each other, and the toes of the two sides sometimes touch.

Internal Features

The diagnostic internal features of Carnegie Stage 21 are

- Corneal body is a compact mesenchymal sheet, two to five cells thick.
- Remnant of ependyma along whole length of the optic stalk, hyaloid groove at bulbar end and a few optic nerve fibers arriving at the brain.
- Cochlea has a return down-curve, i.e., tip turns down.
- Beginning absorption of the thread-like stalk of the hypophysis.
- Oral opening of vomeronasal organ reduced in size, short narrow neck and expanded caudal end.
- Duct of submandibular gland shows simple, stubby primary branching.
- No large glomeruli in metanephros.
- Fibrous zone not very distinct from osteoblast layer.

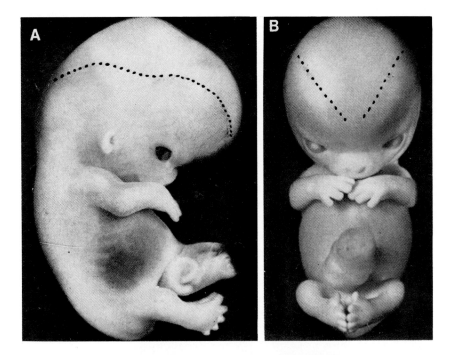

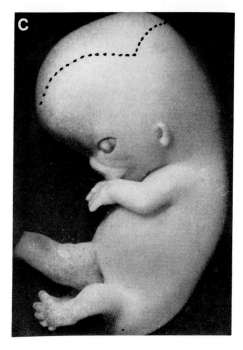

FIGURE 17.   Photographs of three embryos belonging to stage 22. (A) Carnegie embryo No. 6701; 24.0 mm long. (B) Carnegie embryo No. 6832; 25.8 mm long. (C) Carnegie embryo No. 8394; 25.3 mm long. Dotted line indicates advancing edge of vascular plexus. (From Streeter, G. L., *Contrib. Embryol. Carneg. Inst.*, 34, 165, 1951. Carnegie Institution of Washington, Department of Embryology, Davis Division. With permission.)

Man. Carnegie Stage 22

External Features

Embryos of this stage vary from 23.0 to 28.0 mm long and have a postovulatory age of 54 to 56 days. The thickening eyelids are rapidly encroaching upon the eyeballs. The ear is taking shape, and especially the tragus and antitragus have a more definite form. The superficial cranial vascular plexus extends upwards about three-fourths the way above the eye-ear level. The hands extend farther out in front of the body of the embryo and the fingers of one hand may overlap those of the other.

Internal Features

The diagnostic internal features of Carnegie stage 22 are

- A well-defined cuboidal posterior surface layer (Descemet's endothelium) is present. Corneal body sharply layered with large nuclei.
- Sheath layer beginning to form around the optic nerve.
- Transitional stage of cochlea.
- Remnant of incomplete hypophyseal stalk at either end.
- Intermediate stage of vomeronasal organ.
- Secondary branching of submandibular duct, but still mostly solid with suggestion of lumen in distal part.
- Few large glomeruli in metanephros.
- Early shaft shell (or osseus band) of the long bonds is present, but the borders of the shell are not sharp.

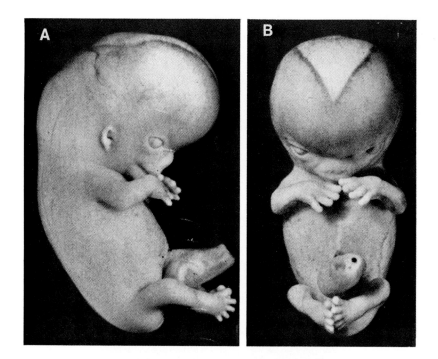

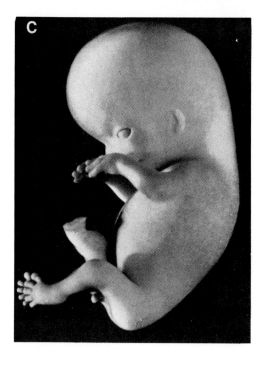

FIGURE 18.   Photographs of two embryos from stage 23. (A and B) Carnegie embryo No. 7425; 27.0 mm long. (C) Carnegie embryo No. 4570; 30.7 mm long. (From Streeter, G. L., *Contrib. Embryol. Carneg. Inst.*, 34, 165, 1951. Carnegie Institution of Washington, Department of Embryology, Davis Division. With permission.)

Man. Carnegie Stage 23

External Features

Embryos of this stage vary from 27.0 to 31.0 mm long and have a postovulatory age of 56 to 60 days. Streeter[9] arbitrarily defined the end of the embryonic period as being marked by invasion of cartilage of the shaft of the humerus by bone and bone marrow cells. It was chosen because it is easily recognized and is of sudden onset. A macroscopic sign appearing at this stage is fusion of the secondary palate, which can be identified by using free hand (razor blade) sections.[4] The head has made rapid progress in its bending towards the erect position. There is a distinct rounding out of the head and a more mature shaping of the neck region and body. The extremities have markedly increased in length and show advancement in the differentiation of their subdivisions. The forearms rise upward to or above the level of the shoulder. The superficial cranial vascular plexus is rapidly approaching the vertex of the head, leaving only a small non-vascular area which will soon become bridged by anastomosing branches.

Internal Features

The diagnostic internal features of Carnegie stage 23 are

- Corneal body thick, fibrous, with thinner elongated nuclei; it is 3 or more times as thick as epithelium.
- Vascular canal present in optic nerve, which now has a definite nerve sheath.
- Tip of cochlea has turned up for second time and may have turned down second time.
- Practically no hypophyseal stalk left, oriented epithelial follicles and abundant angioblasts and capillaries in vascular component of anterior lobe.
- Vomeronasal organ appears as a narrow canal in a long, tapering duct, and regression has begun.
- Submandibular duct is long, much branched; lumen extends into many terminal branches; angiogenesis beginning around epithelium; and mesenchymal capsule forming.
- Numerous large glomeruli with short or long tubules which are becoming convoluted in metanephros.
- All five phases of cartilage cells; elongated shaft shell forming constriction cuff.

Table 4

CARNEGIE STAGES 1 TO 8 OF BABOON EMBRYOS

Carnegie stage	Age (days)	Features
1	1	Fertilization
2	2—5	From 4 to 63 cells
3	5—8	Free blastocyst
4	9 ± 1	Implanting blastocyst
5	10 ± 1	Implanted blastocyst
6	11—15	Primitive chorionic villi, bilaminar embryonic disc
7	16—18	Branching chorionic villi, primitive streak
8	17—19	Primitive groove, beginning notochordal canal in older embryos

Based on Hendrickx, A. G., *Embryology of the Baboon,* University of Chicago Press, Chicago, 1971.

II. BABOON

Carnegie Stages of Baboon Embryos *[Papio cynocephalus]*

In 1971 A. G. Hendrickx[13] and his colleagues published the Embryology of the Baboon which described the external and internal features of 23 stages of baboon development from the fertilized ovum to the end of the embryonic stage. The criteria used for classification of the baboon embryos was almost identical to those used by Streeter[6,10] for human embryos. Comparison of stages 1 to 8 of human and baboon embryos is rendered difficult by their different mode of implantation and placentation and, as with the human embryos, these early stages will merely be listed (Table 4).

In both species somite formation begins in stage 9, and the embryonic period ends in stage 23, which is identified by invasion of the humerus by bone marrow and by fusion of the secondary palate. Between stages 9 and 18 the embryos of man and baboon closely resemble each other, and the various external features appear in the same chronological order. However, from stage 19 onwards, species-specific differences in form begin to appear, and by stage 23 the baboon and human embryos are distinctly different. Baboon and human embryos are approximately the same size up to stage 15, but thereafter baboon embryos are smaller than their human counterparts. At stage 23 baboon embryos range from 25.0 to 28.0 mm, whereas human embryos range from 27.0 to 31.0 mm.

There is an equally exact correlation between the stages at which the various internal features appear in baboon and man. With one exception, there are only minor differences which are no greater than the variation between members of any given stage. In human embryos the vomeronasal organ is a useful diagnostic pointer in stages 19 to 23, where it forms a long tapering duct that expands caudally into a blind tubular sac. In the baboon it is at best a thickening in the nasal septal epithelium with a shallow groove or depression on the surface near its center.

The age of the baboon embryos is the estimated fertilization age (EFA). Full details of the calculation of the EFA and the insemination and minimum age are given in Hendrickx.[13] Stage 23 baboon embryos have an EFA of 47 ± 1 days, whereas the latest figure given for the age of stage 23 human embryos is 56 to 60 days.[2]

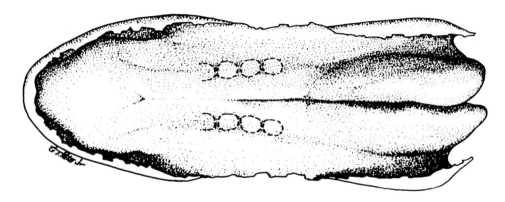

FIGURE 19.   Drawing taken from a reconstruction of a 23-day, three-somite, Carnegie stage 9 baboon embryo. (From Hendrickx, A. G., *Embryology of the Baboon,* University of Chicago Press, Chicago, 1971. With permission.)

Baboon. Carnegie Stage 9

Embryos of stage 9 vary from 1.0 to 2.0 mm in length and have an EFA of 23 ± 1 days. Two of the five embryos described have three pairs of somites. Seen from above, the embryonic shield appears as a pear-shaped disc which is expanding into the amniotic cavity. Prominent neural folds arise from the cranial two thirds of the embryo and caudally fade out into the region of the primitive pit. The older embryos have a cranial (or mesencephalic) flexure.

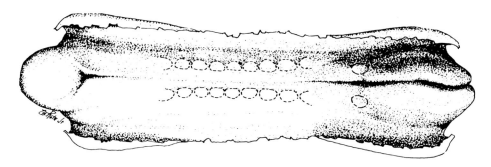

FIGURE 20. Drawing taken from a reconstruction of a 23-day, eight-somite, Carnegie stage 9 baboon embryo. (From Hendrickx, A. G., *Embryology of the Baboon,* University of Chicago Press, Chicago, 1971. With permission.)

Baboon. Carnegie Stage 10

Embryos of stage 10 vary from 2.0 to 3.5 mm in length and have an EFA of 25 ± 1 days. They have from 4 to 12 pairs of somites. This stage is marked by the commencement of the fusion of the neural folds between the 4th and 8th somites to form the neural tube. The embryo changes from a relatively flat plate to a somewhat cylindrical, elongated structure. A "lordosis" similar to that seen in Carnegie stage 9 human embryos may be present. The mandibular and hyoid bars appear as the 7th and 8th somites develop, respectively. The bulge of the heart becomes prominent at the end of this period. The otic placode first appears as a plate of slightly thickened epithelium which begins to invaginate in the older specimens.

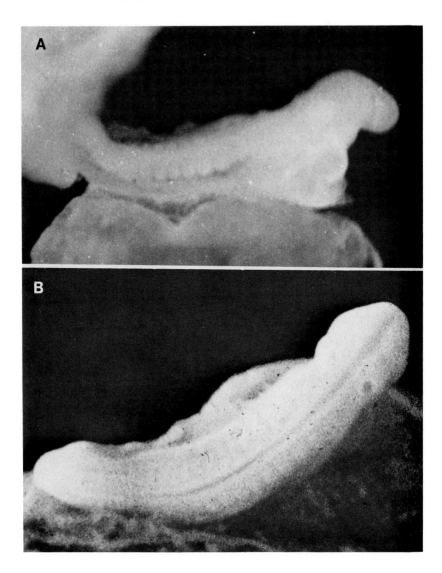

FIGURE 21.   (A) Right aspect of a 27-day, 15-somite, Carnegie stage 11 baboon embryo. (B) Dorsal view of a 27-day, 19-somite Carnegie stage 11 baboon embryo with its caudal end lying parallel to the placenta. (From Hendrickx, A. G., *Embryology of the Baboon*, University of Chicago Press, Chicago, 1971. With permission.)

Baboon. Carnegie Stage 11

Embryos of stage 11 vary from 2.0 to 4.5 mm in length and have an EFA of 27 ± 1 days. They have from 13 to 20 pairs of somites. This is the period of the formation and closure of the cranial neuropore. The mandibular and hyoid bars are well defined. The otic placodes are invaginating to form the otic pits. The tail fold is now formed, and the embryo is elevated above the yolk sac. There is some degree of "lordosis" in all stage 11 embryos, and there is still a broad communication between the yolk sac, foregut, and hindgut.

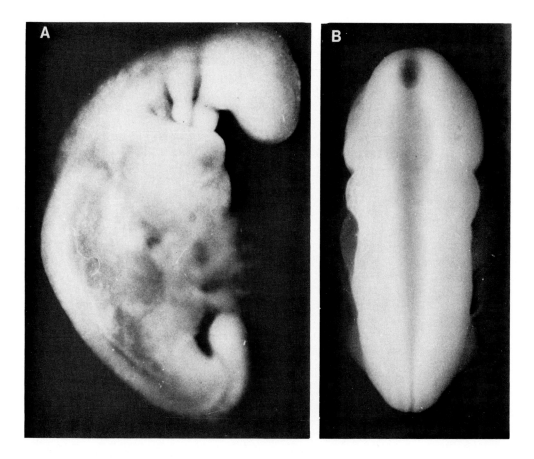

FIGURE 22. (A) Right aspect of a Carnegie stage 12 baboon embryo. (B) Dorsal view of a Carnegie stage 12 baboon embryo. (From Hendrickx, A. G., *Embryology of the Baboon,* University of Chicago Press, Chicago, 1971. With permission.)

Baboon. Carnegie Stage 12

Embryos of stage 12 vary from 3.0 to 4.5 mm in length and have an EFA of 28 ± 1 days. They have 21 to 29 pairs of somites. Their C-shaped curvature clearly distinguishes them from the embryos of the previous stage. The dorsal kink of previous stages is no longer seen. There are three well-defined branchial bars. Caudal to the third bar is a slight depression, the cervical sinus. The caudal neuropore is in the process of closing or is closed. The otic pits still have a small opening. The primordium of the cranial limb bud is seen opposite somites eight to thirteen. The heart bulge is very prominent. The vitelline duct is becoming narrowed in the older members of this group. The elongating tail is blunt, rounded, untapered, and directed cranially.

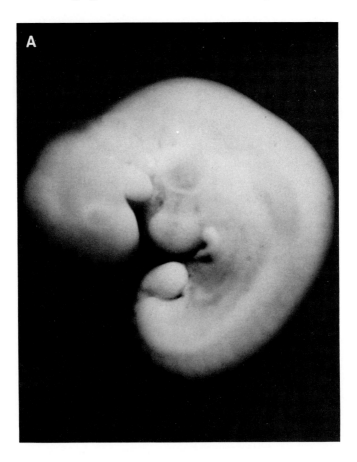

FIGURE 23.   (A) Left aspect of a Carnegie stage 13 baboon embryo. (B) Drawings taken from a reconstruction of the face of a Carnegie stage 13 baboon embryo. (From Hendrickx, A. G., *Embryology of the Baboon*, University of Chicago Press, Chicago, 1971. With permission.)

B

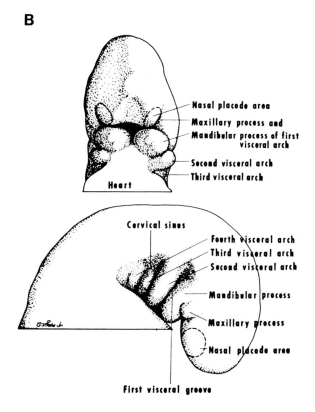

Nasal placode area

Maxillary process and

Mandibular process of first
visceral arch

Second visceral arch

Third visceral arch

Heart

Cervical sinus

Fourth visceral arch

Third visceral arch

Second visceral arch

Mandibular process

Maxillary process

Nasal placode area

First visceral groove

FIGURE 23B.

Baboon. Carnegie Stage 13

Only one stage 13 embryo was obtained. It is 5.6 mm long and has an EFA of 29 ± 1 days. It has 31 pairs of somites. The tip of the cranial limb bud is beginning to turn ventralwards, and the primordium of the hindlimb bud is present. The trunk is more extensively curved, and the upper cervical region is flattened as the cervical curvature develops. The Nackengrube of His is just beginning to appear. Four branchial bars are present, and the first one is dividing into maxillary and mandibular processes. The cervical sinus is clearly visible. The otocyst is now closed. The olfactory placode appears as an ectodermal thickening on the ventrolateral surface of the head.

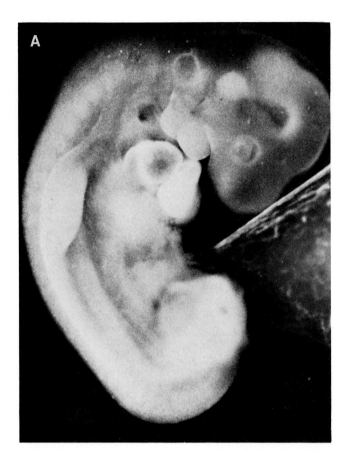

FIGURE 24.   (A) Right view of a Carnegie stage 14 baboon embryo. (B) Drawings taken from a reconstruction of the face of a Carnegie stage 14 baboon embryo. (From Hendrickx, A. G., *Embryology of the Baboon,* University of Chicago Press, Chicago, 1971. With permission.)

B

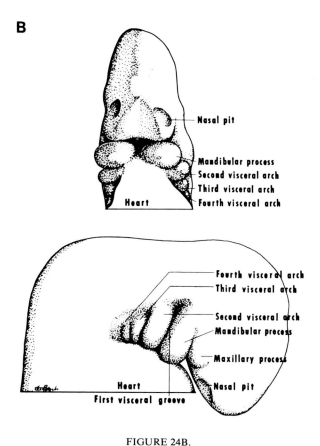

Nasal pit

Mandibular process
Second visceral arch
Third visceral arch
Heart
Fourth visceral arch

Fourth visceral arch
Third visceral arch

Second visceral arch
Mandibular process

Maxillary process

Heart
Nasal pit

First visceral groove

FIGURE 24B.

Baboon. Carnegie Stage 14

Embryos of stage 14 vary from 6.0 to 7.0 mm in length and have an EFA of 30 ± 1 days. The cervical flexure and Nackengrube are better defined. The cerebral hemispheres and mesencephalon are transparent. The cranial limb buds are slightly elongated and curve ventromedially. The smaller caudal limb buds are finlike in the older specimens. Four branchial bars are present. The lens pit is still open. The otocyst is completely separated and appears as a clear vesicle dorsal to the second branchial bar. The endolymphatic duct is seen as a conical elevation of the most dorsal portion of the otocyst. A shallow olfactory pit is visible in the oldest embryo. The tail is longer with an expanded knoblike tip.

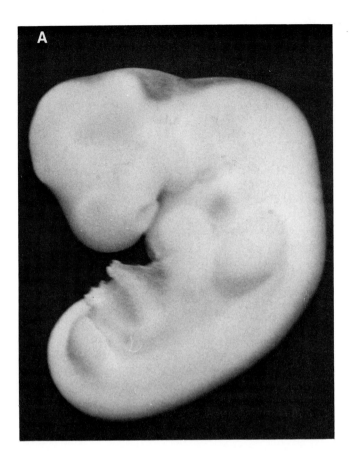

FIGURE 25.   (A) Left view of an older Carnegie stage 15 baboon embryo. (B) Drawings taken from a reconstruction of the face of a Carnegie stage 15 baboon embryo. (From Hendrickx, A. G., *Embryology of the Baboon*, University of Chicago Press, Chicago, 1971. With permission.)

B

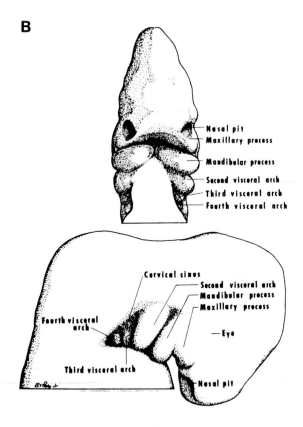

Nasal pit
Maxillary process

Mandibular process

Second visceral arch
Third visceral arch
Fourth visceral arch

Cervical sinus
Second visceral arch
Mandibular process
Maxillary process

Fourth visceral arch

Eye

Third visceral arch

Nasal pit

FIGURE 25B.

Baboon. Carnegie Stage 15

Embryos of stage 15 vary from 6.0 to 8.0 mm in length with an EFA of 31 ± 1 days. Both lens vesicles are completely closed. The olfactory placodes form large oval depressions facing laterally. The limb buds are elongated, and the older specimens show faint constrictions marking the beginning delineation of the hand and footplates. The maxillary process is short and inconspicuous, and the cervical sinus is deepened. The auricular hillocks are appearing on the mandibular and hyoid bars.

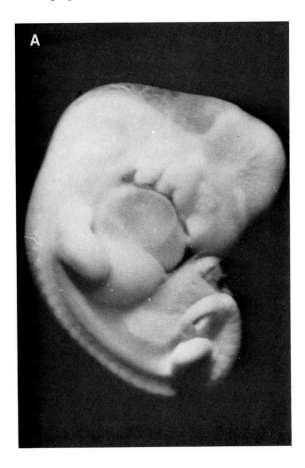

FIGURE 26.   (A) Right view of a Carnegie stage 16 baboon embryo. (B) Drawings taken from a reconstruction of the face of a Carnegie stage 16 baboon embryo. (From Hendrickx, A. G., *Embryology of the Baboon,* University of Chicago Press, Chicago, 1971. With permission.)

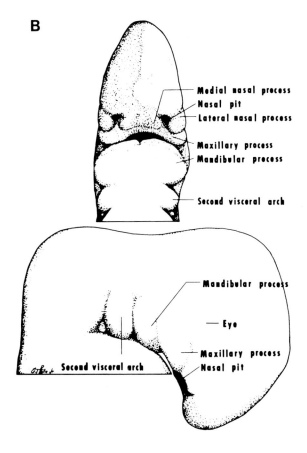

B

Medial nasal process
Nasal pit
Lateral nasal process
Maxillary process
Mandibular process
Second visceral arch

Mandibular process
Eye
Maxillary process
Nasal pit
Second visceral arch

FIGURE 26B.

Baboon. Carnegie Stage 16

Embryos of stage 16 vary from 7.0 to 9.0 mm in length with an EFA of 33 ± 1 days. The nasal pit faces ventrally and is no longer visible in profile view, because it is bounded laterally by the lateral nasal process. The maxillary bar is longer than in the previous stage and forms a prominent surface ridge when seen in profile. The ventral ends of the mandibular and maxillary bars are now inconspicuous. The distal segment of the arm bud is beginning to form a crescentic flange, the primordium of the digital plate. In older specimens the leg bud is divided into a distal foot segment and a proximal leg-thigh segment.

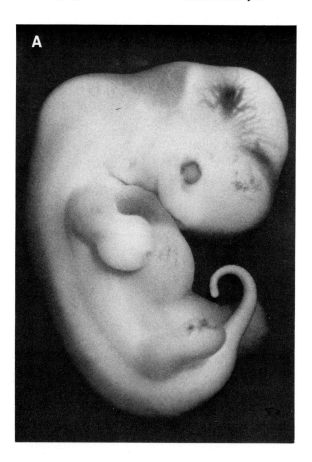

FIGURE 27.   (A) Right view of a Carnegie stage 17 baboon embryo. (B) Drawings taken from a reconstruction of the face of a Carnegie stage 17 baboon embryo. (From Hendrickx, A. G., *Embryology of the Baboon,* University of Chicago Press, Chicago, 1971. With permission.)

B

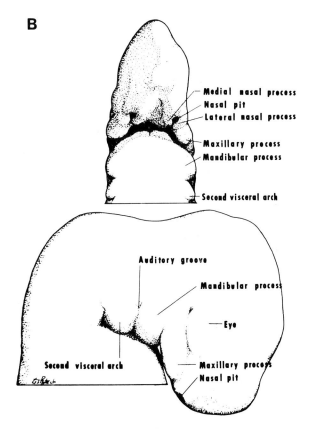

Medial nasal process
Nasal pit
Lateral nasal process
Maxillary process
Mandibular process
Second visceral arch

Auditory groove
Mandibular process
Eye
Second visceral arch
Maxillary process
Nasal pit

FIGURE 27B.

Baboon. Carnegie Stage 17

Embryos of stage 17 vary from 10.0 to 13.0 mm in length with an EFA of 35 ± 1 days. The trunk of stage 17 embryos is straighter than that of the previous stage, but the cervical flexure is still acute. The auricular hillocks form six circumscribed elevations, numbers 1 to 3 on the mandibular bar and 4 to 6 on the hyoid bar. The primordium of the external auditory meatus appears as a shallow depression between the hyoid and mandibular bars. In the hand, the finger rays are becoming visible and the edge of the plate shows signs of crenation. The leg bud has a rounded footplate and leg and thigh regions. The first indication of a pelvic girdle appears as a condensed mass at the junction of leg and trunk. Only the lumbosacral and caudal somites are clearly visible.

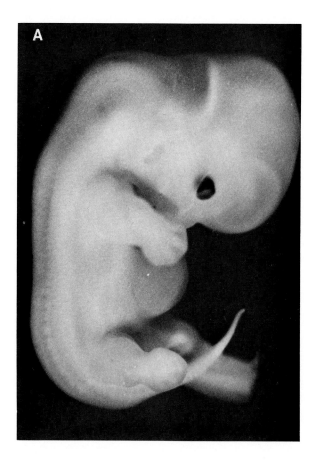

FIGURE 28.   (A) Right view of a Carnegie stage 18 baboon embryo. (B) Drawings taken from a reconstruction of the face of a Carnegie stage 18 baboon embryo. (From Hendrickx, A. G., *Embryology of the Baboon,* University of Chicago Press, Chicago, 1971. With permission.)

B

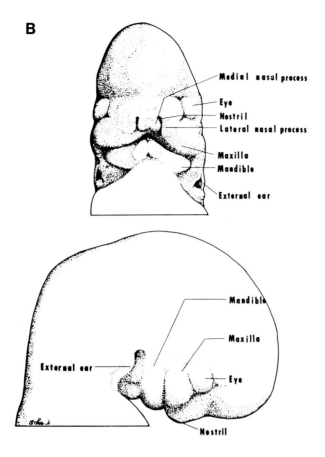

FIGURE 28B.

Baboon. Carnegie Stage 18

Embryos of stage 18 vary from 14.0 to 17.0 mm in length and have an EFA of 37 ± 1 days. Their trunk is straighter and as a result, the head is beginning to move upward and so decreasing the cervical angle. In some specimens the upper and lower eyelids as well as the tip of nose are appearing. The auricular hillocks are prominent and are merging to form the auricle. Distinct finger rays and digital notches are present, and the elbow is recognizable in older embryos. Toe rays are present, but the footplate is not notched. The axes of both arms and legs are almost at right angles to the dorsal line of the body, and they are rotated in a medial direction; hence the palmar surfaces face mediocaudally, whereas the plantar surfaces face medially and slightly cranially. The tail is now relatively straight and tapers to a fine point rather than ending in a blunt knob.

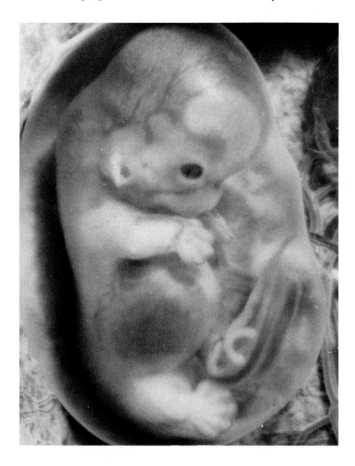

FIGURE 29.   Right view of a Carnegie stage 19 baboon embryo. (From Hendrickx, A. G., *Embryology of the Baboon,* University of Chicago Press, Chicago, 1971. With permission.)

Baboon. Carnegie Stage 19

Embryos of stage 19 vary from 16.0 to 17.0 mm in length with an EFA of 39 ± 1 days. During the final embryonic stages, many significant external features develop which clearly distinguish the baboon embryo from the human embryo. The upper jaw protrudes beyond the plane of the forehead. The upper and lower eyelids are narrow folds at the margins of the eye. The auricular hillocks are coalescing and are less distinct. Epithelial plugs appear in the external nares that are well formed and open laterally. As the back straightens, the cervical angle becomes less acute. The arm, forearm, and hand are delineated. Interdigital notches are distinct on the hand but are just beginning on the foot.

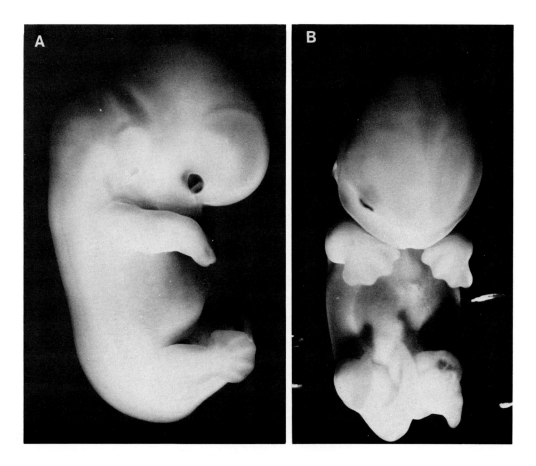

FIGURE 30. (A and B) Right and ventral views of a Carnegie stage 20 embryo. (From Hendrickx, A. G., *Embryology of the Baboon*, University of Chicago Press, Chicago, 1971. With permission.)

Baboon. Carnegie Stage 20

Embryos of stage 20 vary from 17.0 to 18.0 mm in length with an EFA of 41 ± 1 days. The face has a smoother contour than specimens of the previous stage because of blending of the maxilla and premaxilla. The auricular hillocks have merged into a single mass surrounding the external auditory meatus. The limbs have increased in length, and the bends at the elbows and wrists are more pronounced. The short fingers are spread lateralwards. The foot has interdigital notches. Both the thumb and great toe are distinguishable as opposable digits.

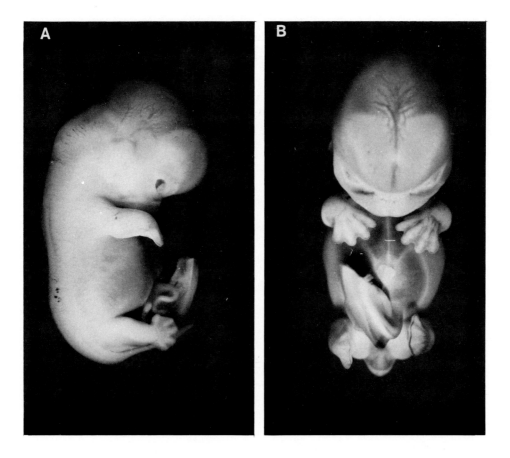

FIGURE 31.   (A and B) Right and ventral views of a Carnegie stage 21 baboon embryo. (From Hendrickx, A. G., *Embryology of the Baboon,* University of Chicago Press, Chicago, 1971. With permission.)

Baboon. Carnegie Stage 21

Embryos of stage 21 vary from 18.0 to 21.0 mm in length and have an EFA of 43 ± 1 days. The face continues to lengthen. The upper and lower jaws elongate at the same rate, but the upper jaw remains further extended. With decrease of the interdigital tissues, the fingers become longer and closer together. The toes become more evident as the interdigital notches of the foot deepen. The vascular plexus at the periphery of the hand and foot is obvious.

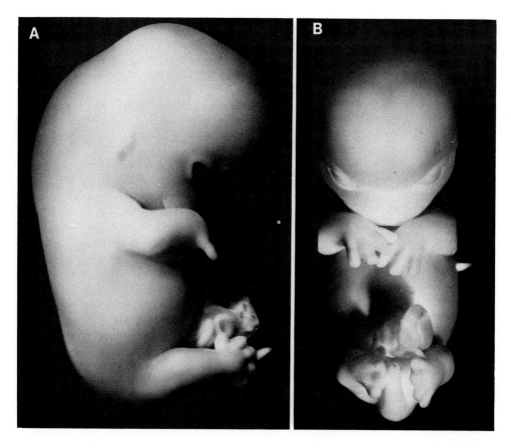

FIGURE 32. (A and B) Right and ventral views of a Carnegie stage 22 baboon embryo. (From Hendrickx, A. G., *Embryology of the Baboon*, University of Chicago Press, Chicago, 1971. With permission.)

Baboon. Carnegie Stage 22

Embryos of stage 22 vary from 21.0 to 23.0 mm in length and have an EFA of 45 ± 1 days. Jaw protrusion is prominent, with the lower jaw growing more rapidly than the upper jaw. The eyelids thicken, and the tragus and antitragus are apparent on a more definitive auricle. The hands extend further out in front of the body, and the fingers of one hand may overlap those on the other side. The feet move closer together. Touch pads appear on the fingertips.

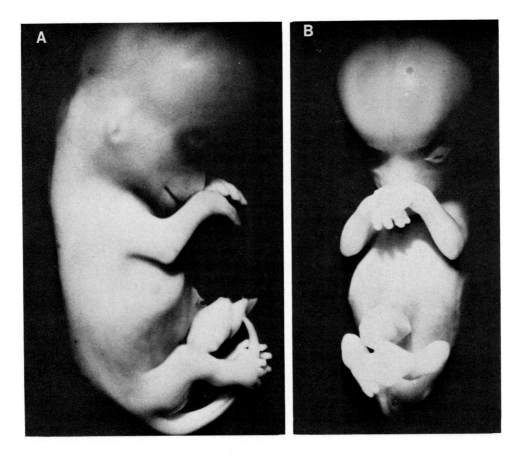

FIGURE 33.   (A and B) Right and ventral views of a Carnegie stage 23 baboon embryo. (From Hendrickx, A. G., *Embryology of the Baboon,* University of Chicago Press, Chicago, 1971. With permission.)

Baboon. Carnegie Stage 23

Embryos of stage 23 vary from 25.0 to 28.0 mm in length and have an EFA of 47 ± 1 days. With extension of the neck, the chin is raised away from the ventral aspect of the chest. The lower jaw protrudes almost as far as the upper jaw. The eyelids cover most of the eye. The auricle is assuming its definitive shape. There is a marked increase in the length of the extremities. The forearm is raised along with the head so that it is above the shoulder. The hands overlap in front of the snout, and the palms face caudalwards. The feet are close to each other and are separated by a long tapering tail that reaches the umbilical cord. Sometimes the feet overlap. The plantar surface of the foot is beginning to turn caudally.

III. RHESUS MONKEY

Carnegie Stages of Rhesus Monkey Embryos *[Macaca mulatta]*

In 1975, Hendrickx and Sawyer[14] published an atlas of rhesus monkey embryos arranged in Carnegie stages 1 to 23. The material illustrated was taken from Heuser and Streeter[15] and the California Primate Research Center. Gribnau and Geijsberts[16] described stages 13 to 23 in considerably greater detail. Comparison of stages 1 to 8 human and rhesus monkey embryos is rendered difficult because of their different mode of implantation and placentation, so these early stages will be merely listed (Table 5).

As in man and baboon, somite formation begins in stage 9, and the embryonic period ends in stage 23 with closure of the secondary palate. Neither Hendrickx and Sawyer[14] nor Gribnau and Geijsberts[16] mention invasion of the humerus by bone marrow. From about stage 19 or 20 onwards, the rhesus monkey embryos begin to differ noticeably in their external form compared to their human counterparts. Up to stage 15 rhesus monkey and human embryos are approximately the same size, but thereafter rhesus monkey embryos lag behind their human counterparts. At the end of stage 23 rhesus monkey embryos range from 24.0 to 30.0 mm, whereas human embryos range from 27.0 to 31.0 mm.

Once again there is a remarkably exact correlation between the stages at which the various internal features appear. The major difference concerns the vomeronasal organ of the rhesus monkey embryo, which is a transient shallow groove like that seen in the baboon embryo.

Hendrickx and Sawyer[14] record the estimated postovulation age, but Gribnau and Geijsberts[16] use the postconceptional age. This was calculated by counting the number of days that had elapsed after the second day of mating. The latter day was regarded as postconceptional day one or embryonic day one. The postovulation age for stages 13 to 23 ranged from 28 to 46 days.[14] The postconceptional age for stages 13 to 23 ranged from 28 ± 1 to 50 ± 1 days.[16]

Table 5
CARNEGIE STAGES 1 TO 8 OF RHESUS
MONKEY EMBRYOS

Carnegie stage	Age (days)	Features
1	1	Fertilization
2	2—4	From 2 to 16 cells
3	5—8	Free blastocyst
4	8—10	Attaching blastocyst
5	10—11	Previllous stage
6	12—15	Primitive villi, early primitive streak
7	16—18	Notochordal process
8	19—20	Primitive pit and neurenteric canal

Based on Hendrickx, A. G. and Sawyer, R. H., Embryology of the rhesus monkey, in *The Rhesus Monkey,* Bourne, G. H., Ed., Academic Press, New York, 1975, 141.

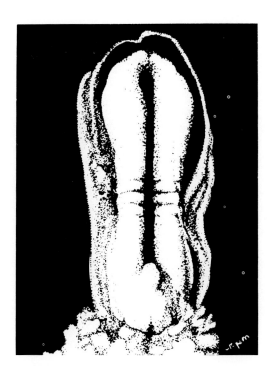

FIGURE 34. Dorsal view of a Carnegie stage 9 rhesus monkey embryo. (From Hendrickx, A. G. and Sawyer, R. H., *The Rhesus Monkey,* Bourne, G. H., Ed., Academic Press, New York, 1975, 141. With permission.)

Rhesus Monkey. Carnegie Stage 9

Embryos of stage 9 are about 2.0 mm long and have a postovulation age of 20 to 21 days. They have up to three pairs of somites. Head and tail folds are present. The neural folds and groove are distinct, and the primitive streak extends from the cloacal membrane to the neurenteric canal and occupies one-quarter to one-third of the length of the embryo. The notochord is prominent.

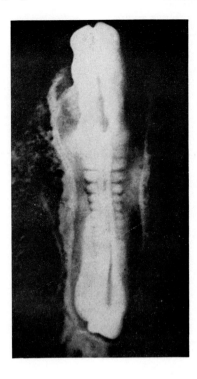

FIGURE 35.   Dorsal veiw of a Carnegie stage 10 rhesus monkey embryo. (From Heuser, C. H. and Streeter, G. L., *Contrib. Embryol. Carneg. Inst.*, 29, 15, 1941. Carnegie Institution of Washington, Department of Embryology, Davis Division. With permission.)

Rhesus Monkey. Carnegie Stage 10

Embryos of stage 10 vary from 2.3 to 3.2 mm in length and have a postovulation age of 21 to 23 days. They have four to 12 pairs of somites. The neural folds are beginning to fuse to form the neural tube. The otic disc is beginning to invaginate. Mandibular and maxillary bars are present.

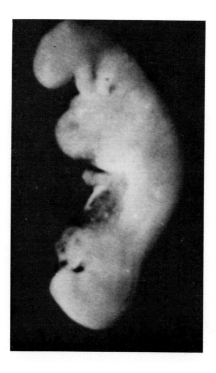

FIGURE 36.   Left view of a Carnegie State 11 rhesus monkey embryo. (From Heuser, C. H. and Streeter, G. L., *Contrib. Embryol. Carneg. Inst.*, 29, 15, 1941. Carnegie Institution of Washington, Department of Embryology, Davis Division. With permission.)

Rhesus Monkey. Carnegie Stage 11

Embryos of stage 11 vary between 2.0 to 3.5 mm in length and have a postovulation age of 24 to 26 days. They have 13 to 20 pairs of somites. The cranial neuropore is closed. The S-shaped heart is prominent. Mandibular and hyoid bars are well defined.

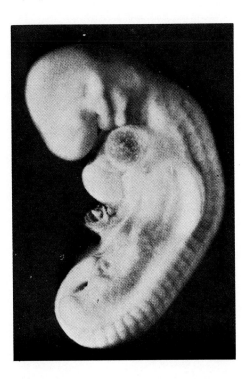

FIGURE 37.   Left view of a Carnegie stage 12 rhesus monkey embryo. (From Heuser, C.
H. and Streeter, G. L., *Contrib. Embryol. Carneg. Inst.,* 29, 15, 1941. Carnegie Institution
of Washington, Department of Embryology, Davis Division. With permission.)

Rhesus Monkey. Carnegie Stage 12

Embryos of stage 12 vary between 3.0 and 5.0 mm in length and have a postovulation age of 27 to 28 days. They have 21 to 29 pairs of somites. The embryo is now C-
shaped, and the dorsal kink of earlier stages is no longer seen. Three branchial bars
are present. The otic pits are closing. The cranial limb bud is forming. The caudal
neuropore is closed. The yolk stalk has formed.

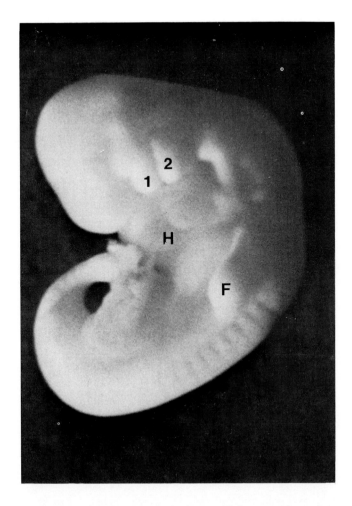

FIGURE 38.   Left view of a Carnegie state 13 rhesus monkey embryo.
1,2 = first and second branchial bars; F = cranial limb bud; H = heart
bulge. (From Gribnau, A. A. M. and Geijsberts, L. G. M., *Adv. Anat.
Embryol. Cell Biol.*, 68, 1, 1981. Springer-Verlag, Heidelberg. With permission.)

Rhesus Monkey. Carnegie Stage 13

Embryos of stage 13 vary between 4.5 to 6.0 mm in length and have a postconceptional age of 28 ± 1 to 30 ± 1 days. A definite cranial limb bud is present, but the
caudal limb bud is only just beginning to appear. The central nervous system is still the
main determinant of body form. The prominent heart bulge is almost in contact with
the head. Three branchial bars are present, the first and second being very prominent.
The first branchial bar is subdividing into maxillary and mandibular parts. The cranially directed, short tail has a knob-like end. A flat olfactory placode is present. The
otic vesicle is completely closed and detached from the surface. There is a slight indentation of the lens placode in the older specimens.

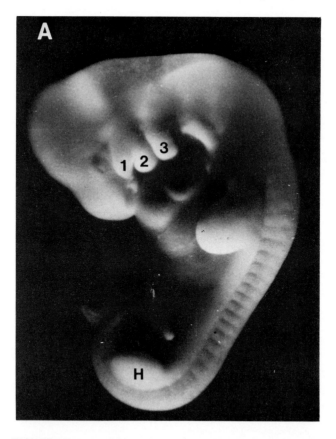

FIGURE 39.   (A and B) Left views of a Carnegie stage 14 rhesus monkey embryo. In B the embryo has been cleared in methylbenzoate. 1 = maxillary process; 2 = mandibular process; 3 = hyoid bar; H = caudal limb bud. (From Gribnau, A. A. M. and Geijsberts, L. G. M., *Adv. Anat. Embryol. Cell Biol.,* 68, 1, 1981. Springer-Verlag, Heidelberg. With permission.)

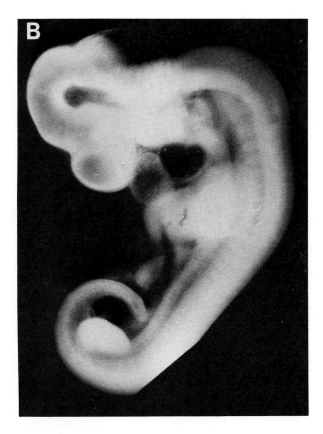

FIGURE 39B.

Rhesus Monkey. Carnegie Stage 14

Embryos of stage 14 vary in length from 6.0 to 8.0 mm and have a postconceptional age of 30 ± 1 to 32 ± 1 days. As before, the contour of the embryo is largely determined by the central nervous system. The somites are very distinctly visible. The various subdivisions of the heart can be seen from the outside. The maxillary process is now much more distinct. Both a deep olfactory pit and lens pit are visible. The enlarged cranial limb is beginning to show separation into proximal and distal parts. A distinct caudal limb bud is present. The tail is lengthened and now curves around the caudal limb bud, its tip pointing dorsally. There is a well-defined endolymphatic duct.

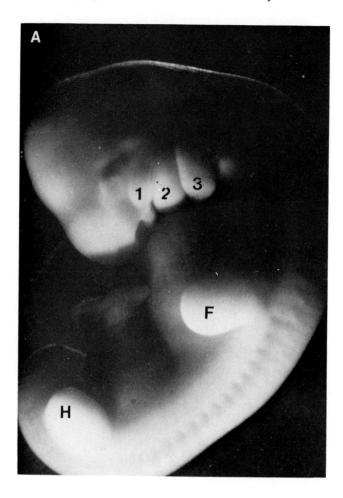

FIGURE 40. (A and B) Left views of a Carnegie stage 15 rhesus monkey embryo. In B the embryo has been cleared in methylbenzoate. 1 = maxillary process; 2 = mandibular process; 3 = hyoid bar; F = cranial limb bud; H = caudal limb bud. (From Gribnau, A. A. M. and Geijsberts, L. G. M., *Adv. Anat. Embryol. Cell Biol.*, 68, 1, 1981. Springer-Verlag, Heidelberg. With permission.)

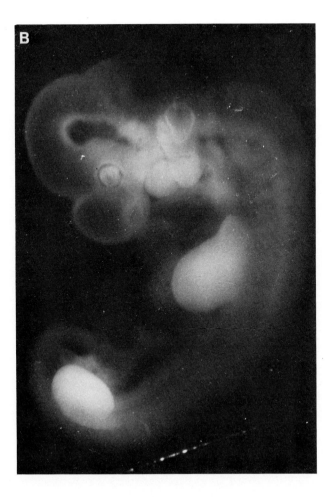

FIGURE 40B.

Rhesus Monkey. Carnegie Stage 15

Embryos of stage 15 vary in length between 7.0 and 9.0 mm and have a postconceptional age of 30 ± 1 to 33 ± 1 days. The C-shaped body axis is still dominated by the cervical flexure. The somites are prominent structures from the cervical to the coccygeal region. The maxillary process has enlarged considerably but is separated from the lateral nasal process by a deep nasolacrimal cleft. Auricular hillocks are beginning to appear on the hyoid bar. The deep oval olfactory pit is located somewhat laterally. Both lens vesicles are usually closed. The cranial limb bud is subdivided into a proximal arm segment and a distal hand segment which is directed medially. The caudal limb bud is lengthened but not yet subdivided. The tail has lengthened and, in some advanced specimens, has a terminal knob.

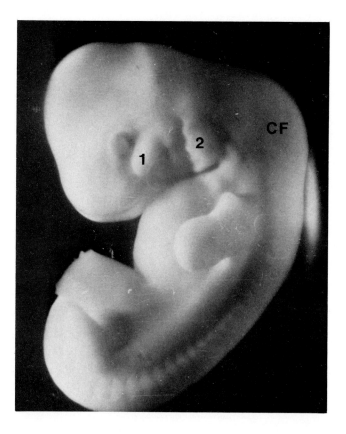

A

FIGURE 41.   (A and B) Left views of a Carnegie stage 16 rhesus monkey
embryo. In B the embryo has been cleared in methylbenzoate. 1 = maxillary process; 2 = hyoid bar; CF = cervical flexure; Cr = cranial flexure.
(From Gribnau, A. A. M. and Geijsberts, L. G. M., *Adv. Anat. Embryol. Cell Biol.*, 68, 1, 1981. Springer-Verlag, Heidelberg. With permission.)

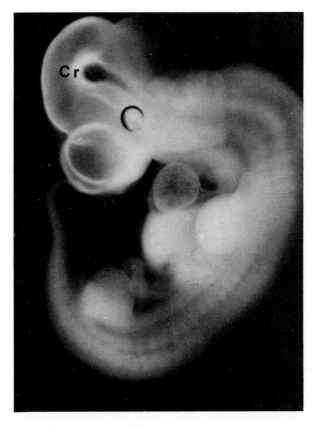

FIGURE 41B.

Rhesus Monkey. Carnegie Stage 16

Embryos of stage 16 vary in length from 7.0 to 11.0 mm and have a postconceptional age of 32 ± 1 to 34 ± 1 days. The nasal pits have a more medial position than in the preceding stage. Retinal pigment is faintly recognizable externally but is clearly seen if the embryo is cleared in methylbenzoate. There is a definite rounded handplate, but only the extended caudal limb bud of older specimens shows the first sign of demarcation of the footplate. There are distinct auricular hillocks on the hyoid bar. The somites are no longer visible in the cervical region. The tail is lengthened and may have a knob-like end. The lens vesicles are invariably closed.

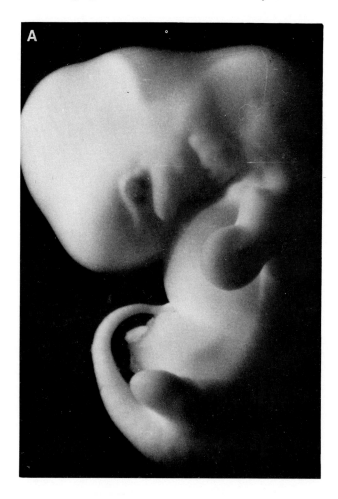

FIGURE 42.   (A and B) Left views of a Carnegie stage 17 rhesus monkey embryo. In B the embryo has been cleared in methylbenzoate. (From Gribnau, A. A. M. and Geijsberts, L. G. M., *Adv. Anat. Embryol. Cell Biol.*, 68, 1, 1981. Springer-Verlag, Heidelberg. With permission.)

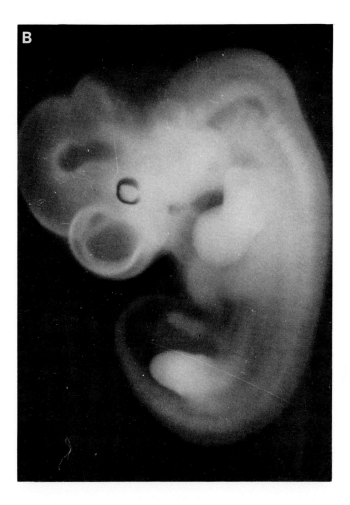

FIGURE 42B.

Rhesus Monkey. Carnegie Stage 17

Embryos of stage 17 vary in length from 9.0 to 12.0 mm and have a postconceptional age of 34 ± 1 to 36 ± 1 days. The somites are less conspicuous than in the previous stage. The nasomaxillary and nasolacrimal grooves have formed. The deep nasal pits open more medially and the primitive nostrils are hardly visible in profile views. Auricular hillocks are now seen on both the mandibular and hyoid bars. The handplate is subdivided into a central convex carpal plate and a peripheral flat fingerplate. In the older members the first sign of finger rays is seen. A well-marked tubercle marks the boundary between the forearm and arm segments of the forelimb. A true footplate is seen in the older members. The tail has increased considerably in length, and its tip may even reach up to the level of the forehead.

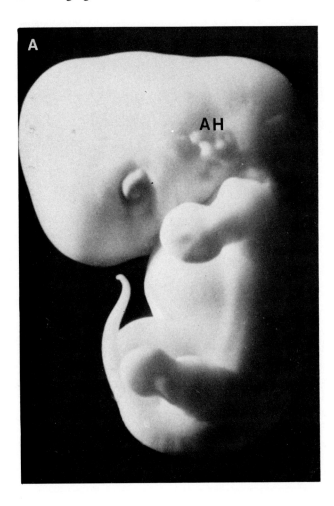

FIGURE 43.   (A and B) Left views of a Carnegie stage 18 rhesus monkey embryo. In B the embryo has been cleared in methylbenzoate. AH = auricular hillocks; PF = pontine flexure. (From Gribnau, A. A. M. and Geijsberts, L. G. M., *Adv. Anat. Embryol. Cell Biol.*, 68, 1, 1981. Springer-Verlag, Heidelberg. With permission.)

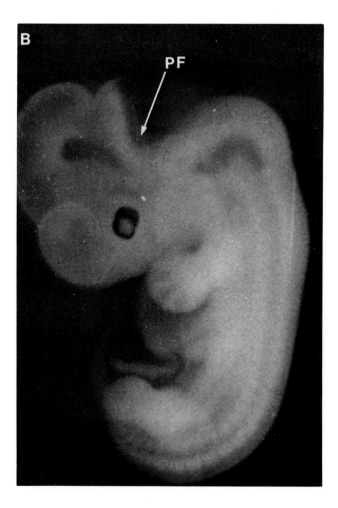

FIGURE 43B.

Rhesus Monkey. Carnegie Stage 18

Embryos of stage 18 vary in length from 11.0 to 15.0 mm and have a postconceptional age of 35 ± 1 to 38 ± 1 days. The head has begun to lift, and the cervical flexure is now a right angle. Somites are only recognizable in the lumbosacral region. The maxillary process is less protruding. The heavy retinal pigmentation is very obvious. The primitive external auditory meatus is surrounded by six very prominent auricular hillocks. There are distinct finger rays in the handplate, and in the older members it has a crenated edge because of the developing interdigital notches. The palmar surfaces of the handplates are facing each other, and in the older ones they have begun to face caudally as well. The rounded and flattened footplates do not show any toe rays. Their plantar surfaces are facing mediocranially. The tail is still of considerable length and tapers into a fine curved tip.

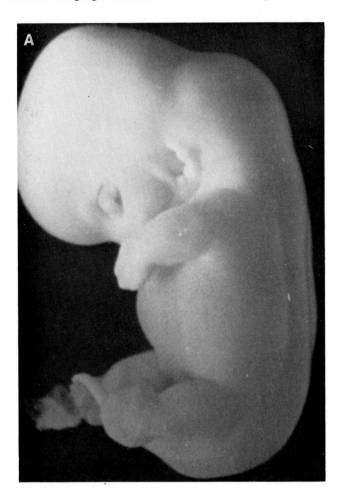

FIGURE 44.   (A) Left view of a Carnegie stage 19 rhesus monkey embryo. (B) Ventral view of the same embryo cleared in methylbenzoate.
(From Gribnau, A. A. M. and Geijsberts, L. G. M., *Adv. Anat. Embryol. Cell Biol.*, 68, 1, 1981. Springer-Verlag, Heidelberg. With permission.)

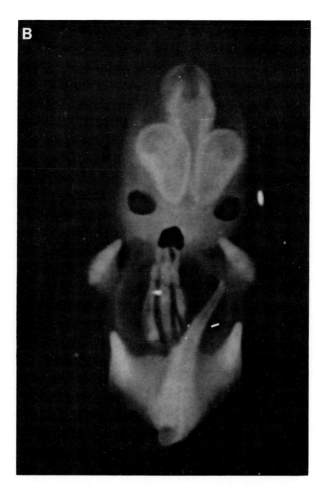

FIGURE 44B.

Rhesus Monkey. Carnegie Stage 19

Embryos of stage 19 vary in length from 14.0 to 17.0 mm and have a postconceptional age of 36 ± 1 to 42 ± 1 days. The head has lifted further, and the cervical flexure is more than a right angle. The cervical flexure of the brain forms a distinct bulge. The spinal cord protrudes as a longitudinal column, and the somites are no longer visible. The folds of the eyelids are appearing. The six auricular hillocks are still recognizable. The pentagonal footplates have definite toe rays. Their plantar surfaces now face medially. The tail is relatively shorter and only reaches up the level of the unbilicus.

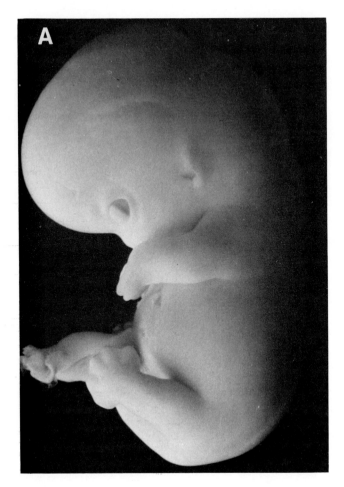

FIGURE 45.   (A and B) Left and ventral views of a Carnegie stage 20 rhesus monkey embryo. (From Gribnau, A. A. M. and Geijsberts, L. G. M., *Adv. Anat. Embryol. Cell Biol.*, 68, 1, 1981. Springer-Verlag, Heidelberg. With permission.)

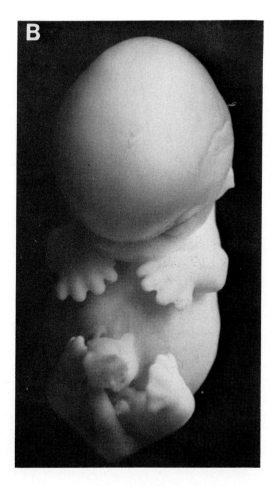

FIGURE 45B.

Rhesus Monkey. Carnegie Stage 20

Embryos of stage 20 vary in length from 16.0 to 20.0 mm and have a postconceptional age of 38 ± 1 to 42 ± 1 days. Progressive lifting of the head has increased the cervical flexure to about 120°. The bulge of the cervical flexure and the spinal cord are less marked than in the previous stage. The head is quadrangular in shape with a slight protrusion of the forehead. The eyelids cover the upper and lower parts of the eyes. Primordial hair follicles can be seen at the future site of the eyebrows. Definite auricles have been formed and are covering up the external auditory meatus. The arms and legs are almost parallel and approximately at right angles to the dorsum of the embryo. Slight flexures indicate the elbow and knee. The hands have rotated further, and their palmar surfaces face caudally. The short, stubby fingers are beginning to be separated. The toe rays are very prominent, and the footplate has a crenated margin. The tail length is about the same as in the preceding stage.

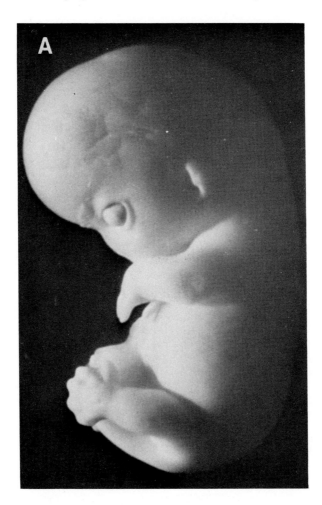

FIGURE 46.   (A and B) Left and ventral views of a Carnegie stage 21 rhesus monkey embryo. In B the embryo has been cleared in methylbenzoate. (From Gribnau, A. A. M. and Geijsberts, L. G. M., *Adv. Anat. Embryol. Cell Biol.*, 68, 1, 1981. Springer-Verlag, Heidelberg. With permission.)

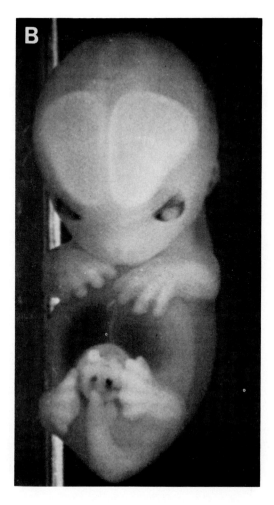

FIGURE 46B.

Rhesus Monkey. Carnegie Stage 21

Embryos of stage 21 vary in length from 18.0 to 22.0 mm and have a postconceptional age of 40 ± 1 to 44 ± 1 days. As the head continues to rise, the neck region becomes apparent in lateral views, and in frontal views the face is clearly visible. The cervical bulge has disappeared, and there is further frontal protrusion of the forehead. The eyelids now cover about one quarter of the external surface of the eyes. The hair follicles of the eyebrows are more distinct. The auricles project laterally, and the primordial helix and antihelix are visible. The jaws have elongated, but the upper jaw extends further forward than the lower. The lengthened arms and legs now have obvious flexures at the elbow and knee. The hands are palmarflexed at the wrist. The distal phalanges are separated. The right and left hands almost touch in front of the snout. The ankles may be recognizable in older specimens. The footplates still face medially but are closer together. The tips of the toes are beginning to separate.

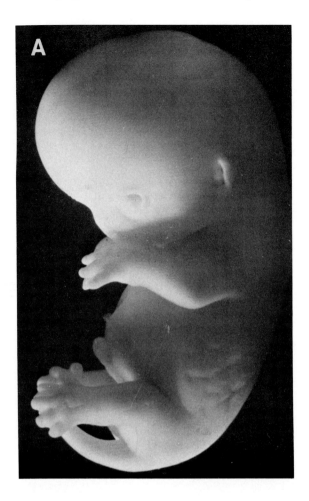

FIGURE 47.   (A and B) Left and ventral views of a Carnegie
stage 22 rhesus monkey embryo. In B the embryo has been cleared
in methylbenzoate. (From Gribnau, A. A. M. and Geijsberts, L.
G. M., *Adv. Anat. Embryol. Cell Biol.,* 68, 1, 1981. Springer-
Verlag, Heidelberg. With permission.)

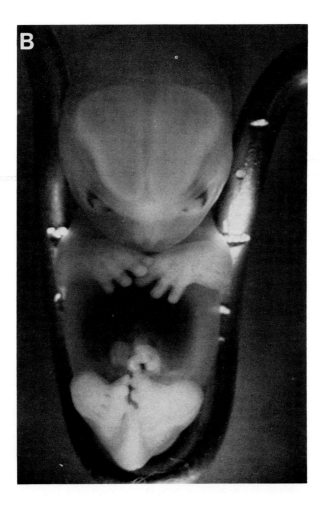

FIGURE 47B.

Rhesus Monkey. Carnegie Stage 22

Embryos of stage 22 vary in length from 20.0 to 25.0 mm and have a postconceptional age of 44 ± 1 to 48 ± days. Elevation of the large head has continued, and the
chin is off the chest. The frontal region of the head is larger than the occipital region.
The upper jaw still extends further forward than the lower jaw. The eyes have a more
frontal situation, and one third of the eyeball is covered by the eyelids. The hair follicles of the eyebrows have extended over the root of the nose. In addition to the helix
and antihelix, the tragus and antitragus are beginning to appear. The upper arms and
legs have noticeably lengthened. The flexures at the knee and elbow have increased,
and the ankles and wrists are easily recognizable. The fingers are separated to about
the proximal phalanges. The fingers of the right and left hands partially overlap. The
plantar surfaces of the feet are parallel to the midsaggital plane, with the tail in between
them. The distal phalanges of the toes are separated, and the toes of the left and right
foot almost touch each other. Both the thumb and great toe are abducted.

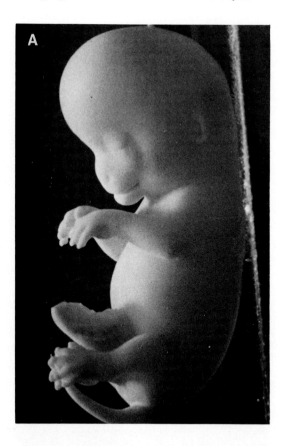

FIGURE 48.   (A and B) Left and ventral views of a Carnegie stage 23 rhesus monkey embryo. (From Gribnau, A. A. M. and Geijsberts, L. G. M., *Adv. Anat. Embryol. Cell Biol.*, 68, 1, 1981. Springer-Verlag, Heidelberg. With permission.)

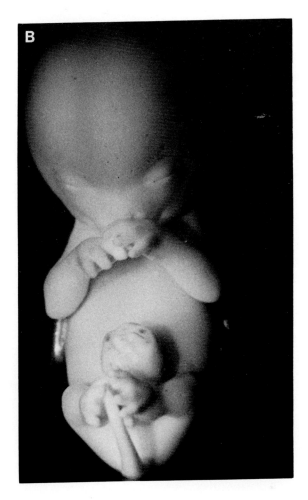

FIGURE 48B.

Rhesus Monkey. Carnegie Stage 23

Embryos of stage 23 vary in length from 24.0 to 30.0 mm and have a postconceptional age of 46 ± 1 to 50 ± 1 days. The neck continues to extend and elevates the head and chin. The volume of the neurocranium is still greater than the splanchnocranium, but the latter is increasing. The low jaw now protrudes almost as far as the upper jaw. The occipital region of the head is relatively larger than before. The eyelids of the older members are completely fused. The auricles have assumed their definitive shape. The limbs are longer and have more pronounced flexures at the elbow and knee. The hands are in front of the snout, with their palmar surfaces facing caudally. The two hands partially overlap. Separation of the fingers is now complete. The position of the feet and tail is similar to the preceding stage. The toes of the right and left foot may touch. Separation of the toes is also completed. The tips of the toes and fingers have a swollen appearance.

IV. COMMON MARMOSET

Carnegie Stages of Common Marmoset Embryos [Callithrix jacchus]

Phillips[17] described 36 marmoset embryos and arranged them into Carnegie stages 7 to 11, 13 to 17, 19 and 21. Marmosets commonly have twins which are enclosed in a common chorion, but each twin has its own discoidal placenta. However, the placental vessels are connected by extensive vascular anastomoses. In the cases of stages 7 and 8, the embryo was not identifiable visually, and the conceptual sac was serially sectioned. Stage 7 is regarded as the beginning of the embryonic period, and the trilaminar embryonic disc has an early primitive streak. The stage 8 embryos have a trilaminar disc with a notochord and primitive groove. The minimum and maximum ages were: stage 7, 25 to 28 days; stage 8, 42 to 49 days. Only one stage 9 embryo was recovered, but it had not yet developed any somites. Its minimum and maximum ages were 55 to 59 days. One set of twin embryos of stage 10 were recovered. They had seven pairs of somites and prominent neural folds. Both had marked cranial and dorsal flexures. Their minimum and maximum ages were 51 to 59 days.

In general, there are only minor differences in the stages at which the various organs develop in the marmoset as compared to man, baboon, and rhesus monkey. The formation of the vomeronasal organ in the marmoset is very similar to that of the human embryo as compared to its rudimentary appearance in the baboon and rhesus monkey.

At all stages of development, marmoset embryos are considerably shorter than their human counterparts; e.g., at stage 21 marmoset embryos range from 14.4 to 15.9 mm compared to 22.0 to 24.0 mm for human embryos. Also, marmoset embryos of any given stage are very much older than comparable human embryos; e.g., at stage 19 marmoset embryos are 75 days old as compared to 48 to 51 days for human embryos.

Marmoset. Carnegie Stage 11

Embryos of stage 11 vary in length from 1.3 to 2.2 mm and have minimum and maximum ages of 52 to 60 days. They have from 15 to 18 pairs of somites. The cranial neuropore is widely open. The mandibular and hyoid bars are well developed. The otic placode is readily visible as an epithelial plate in varying stages of invagination.

Marmoset. Carnegie Stage 13

Embryos of stage 13 vary in length from 2.4 to 4.8 mm and have minimum and maximum ages of 51 to 70 days. All have more than thirty pairs of somites. There are three well-developed branchial bars, and a fourth bar is visible in some specimens. The otic and optic vesicles are well defined and appear as opaque circles with transparent centers. The otocyst is closed. The lens placode is present but has not yet begun to invaginate. The olfactory placode is not visible. The cranial limb bud is present in all specimens, and the caudal limb buds are appearing in some. The tail is half curved and ends in a blunt knob.

Marmoset. Carnegie Stage 14

Embryos of stage 14 vary in length from 4.1 to 5.0 mm and have minimum and maximum ages of 61 to 63 days. They have a prominent cervical flexure, and the Nackengrube of His is present. The cranial limb buds have begun to elongate and curve ventromedially. In some, the caudal limb bud has a more definite fin-like form.

Marmoset. Carnegie Stage 15

Embryos of stage 15 vary in length from 6.8 to 8.3 mm and have minimum and maximum ages of 54 to 75 days. The nasal pits are forming. Four branchial bars are present. The lens vesicles are closed. The elongated cranial limb bud is differentiated into handplate and arm-shoulder segment. The caudal limb bud is less obviously differentiated.

Marmoset.Carnegie Stage 16

Embryos of stage 16 vary in length from 8.3 to 9.2 mm and have minimum and maximum ages of 66 to 83 days. The cervical flexure is not as marked as in human or baboon embryos at this stage. Retinal pigment is visible. The mandibular and hyoid bars are prominent. Somites are clearly visible from just anterior to the forelimb to the tip of the tail. An annular constriction at the wrist marks the rounded handplate from the forearm. Digital rays are not visible. The caudal limb bud is beginning to show differentiation into foot and leg-thigh regions.

Marmoset, Carnegie Stage 17

A stage 17 embryo is 12.0 mm in length and has a minimum and maximum age of 61 to 77 days. The head and neck regions represent nearly one half of the total length of the embryo. The cervical flexure is not as great as that of human or baboon embryos at this stage. The large heart obscures the olfactory region and the auricular hillocks in the intact embryo. The primordium of the external auditory meatus can just be distinguished between the mandibular and hyoid bars. The clearly demarcated, round handplate has digital rays, but its margin is not crenated. There is a distinct rounded footplate. Somites are only visible from the lumbosacral region to the tip of the tail.

Marmoset. Carnegie Stage 19

Embryos of stage 19 vary in length from 12.1 to 12.4 mm (a pair of twins) and have an insemination age of 75 days. The head represents approximately one third of the total body length. The reduced cervical flexure and elongated trunk give the appearance of a distinct neck. The snout is less pronounced than in baboon embryos of this stage. The eye is heavily pigmented. The auricle and external auditory meatus are prominent. The forelimbs are divided into arm-shoulder, forearm, and handplate segments and the hindlimbs into leg-thigh and footplate regions. The palmar surfaces of the hands point caudomedially and the plantar surface of the feet craniomedially. The digital rays and interdigital notches are well marked on the handplate. The footplates have only faint digital rays and no interdigital notches. The tail has lost the terminal knob seen earlier and has a pointed tip.

Marmoset. Carnegie Stage 21

One set of triplets of stage 21 embryos vary in length from 14.4 to 15.8 mm and have a minimum and maximum age of 73 to 87 days. The head is smooth in outline and the cervical flexure and snout are not as pronounced as in baboon embryos of this stage. The heart is still relatively large compared to other primate embryos of this stage. The limbs are larger and more clearly subdivided. The fingers are longer and close together. The toes are less separated than the fingers. The palms of the hands point caudomedially and the soles of the feet craniomedially. The eyelids cover about half the surface of the eye.

V. LESSER GALAGO

Carnegie Stages of Lesser Galago Embryos *[Galago senegalensis]*

Lesser galago embryos resemble marmoset embryos in terms of size and age as contrasted to the embryos of man, rhesus monkey, and baboon. The primitive streak appears at 25 to 28 days in the lesser galago and marmoset but at 17 days in the baboon and rhesus monkey and 16 days in man. By stage 23 the lesser galago embryo is 14.0 mm long, compared to 30.4, 27.5, and 26.5 mm in man, rhesus monkey, and baboon, respectively. Despite these differences, their external and internal appearances follow the same pattern of development as the other primate embryos. The lesser galago, as contrasted to baboon and rhesus monkey, has a well-developed vomeronasal organ very comparable to that seen in the human embryo. At stage 23 the embryos of man, baboon, rhesus monkey, and lesser galago have distinctly different external features. By stage 23, many of the species-specific internal features of the lesser galago are present. It is noteworthy that the characteristically elongated tarsus of the lesser galago is not seen until about midterm.

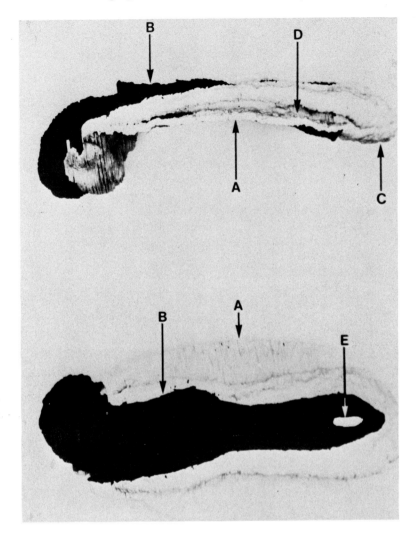

FIGURE 49.   Left lateral (above) and dorsal (below) aspects of a reconstruction of a Carnegie stage 11 lesser galago embryo. A = cut edge of yolk sac; B = cut edge of amnion; C = allantois; D = coelom; E = posterior neuropore.

Lesser Galago. Carnegie Stage 11

An embryo of stage 11 has a greatest length of 3.2 mm, as estimated from the number and thickness of the sections. Its age is unknown. There are 18 pairs of somites. Caudal to the heart, the dorsoventrally flattened body of the embryo has a slight, even, ventral concavity and not the "lordosis" seen in the previous species at this stage. The cranial end of the embryo is flexed ventrally and twisted slightly to the right. The widely open cranial neuropore faces cranially, ventrally, and caudally. The otic placodes form very shallow depressions on the dorsolateral aspect of the head. The mandibular bars are very prominent. There is no tail fold, and the extreme caudal end of the embryo is formed by the allantois. The slit-like caudal neuropore is on the dorsal surface of the embryo just cranial to the root of the allantois. The large, shallow midgut occupies over half the total length of the embryo and opens into the voluminous yolk sac, which is 10.0 mm in diameter.

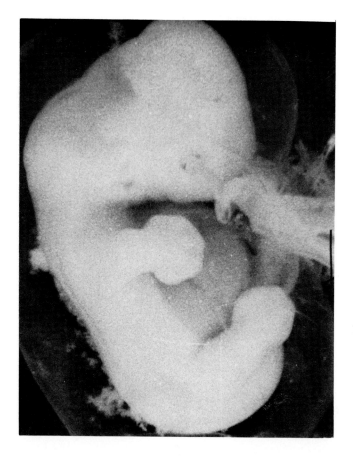

FIGURE 50. Right aspect of Carnegie stage 18 lesser galago embryo.

Lesser Galago. Carnegie Stage 18

An embryo of stage 18 is 8.5 mm long. Its age is unknown. The eye is clearly demarcated by a narrow circular ring of pigment, bounded by a shallow groove caused by the commencement of folds of the eyelids. The auricular hillocks are beginning to fuse. There is a flattened handplate whose margin has shallow interdigital notches. The elbow is not yet visible. The shorter hindlimb has a footplate with no interdigital notches. The long tail curves ventralwards between the hindlimb buds, and its tip lies alongside the left side of the trunk. There is a well-marked cervical flexure. The extreme cranial end of the embryo is formed by the protuberant midbrain.

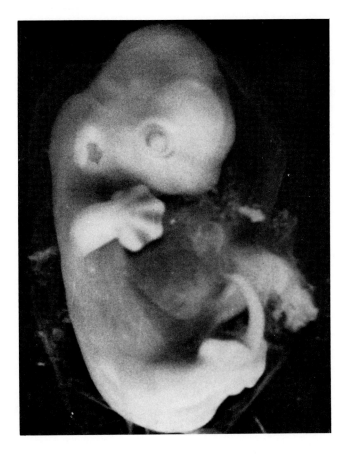

FIGURE 51.   Right aspect of a Carnegie stage 19/20 lesser galago embryo.

Lesser Galago. Carnegie Stage 19/20

An 11.0 mm lesser galago has features of late stage 19 and early stage 20. Its age is unknown. The eye pigmentation has increased, and the folds of the eyelids are more prominent. The auricular hillocks have fused, and the general outline of the ear is seen. The longer limb buds are about equal in length, but the forelimb is more advanced than the hindlimb. The finger rays form distinct elevations separated by deep depressions, and the margin of the handplate is distinctly notched. The footplate is at the same stage as the handplate of the previous embryo. The elbow is just appearing, but the knee is not yet visible. The tail is long and has a pointed tip. The forehead is less protuberant than that of stage 19 human embryos.

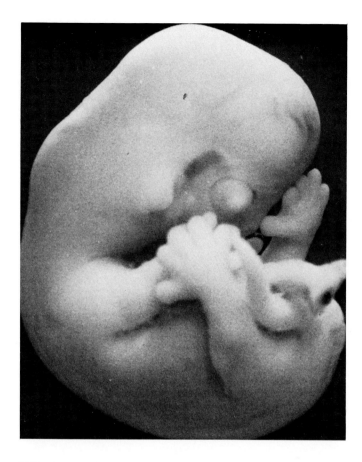

FIGURE 52.   Right aspect of a Carnegie stage 22 lesser galago embryo.

Lesser Galago. Carnegie State 22

A stage 22 lesser galago embryo has an acute flexure of the shoulder region, and so is probably somewhat more than 12.0 mm long. Its age is unknown. The elbow and knee joints are visible, and both hands and feet have distinct digits. There is no elongation of the tarsus, but both the pollex and hallux diverge widely from the other digits. The eyes are more covered by the eyelids, and the pinna of the ear is well developed. The forehead is somewhat more protuberant but not nearly as much as in stage 22 human embryos.

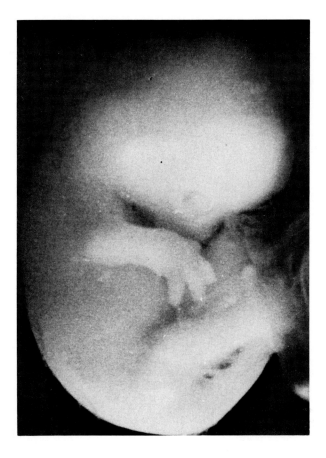

FIGURE 53.   Right aspect of a Carnegie stage 23 lesser galago embryo.

Lesser Galago. Carnegie Stage 23

A stage 23 embryo is 14.0 mm in length, and its age is estimated to be 61 to 62 days postcopulation. It is much smaller than human embryos of this stage, which range in length from 27.0 to 31.0 mm. The external features of this embryo are very much like those of the previous specimen, except that it has a smooth dorsal curvature instead of the acute flexure in the shoulder region. Its extreme anterior end is formed by the apex of the midbrain flexure. The forehead is less protuberant than in its human counterparts. Not only is the head more angular in outline compared to that of a human embryo, but it is relatively smaller. The point score for the eight key organs used by Streeter[10] is 51, so it stands close to early stage 23 human embryos. The secondary palate is closed.

Table 6
CARNEGIE STAGES 1 TO 8 OF MOUSE EMBRYOS

Carnegie stage	Age (days)	Features	Theiler stage
1	1	One-celled egg	1
	2	Segmenting egg	2
2	3	Morula	3
	4	Advanced segmentation	4
3	5	Blastocyst	5
4	5.5	Implantation	6
5	6	Formation of egg cylinder	7
	7	Differentiation of egg cylinder	8
6	7.5	Advanced endometrial reaction	9
7	8	Amnion and primitive streak formation	10
8	8.5	Neural plate, presomite stage	11

Based on Theiler.[18]

VI. MOUSE AND RAT

Carnegie Stages of Mouse Embryos [Mus musculus]

Theiler[18] described 27 stages of development of the house mouse ranging from the fertilized ovum to the newborn. He related his stages 1 to 22 to Carnegie stages 1 to 23. We recognize some errors in correlating the two staging systems and have reclassified certain of Theiler's stages in relation to Carnegie stages. This reclassification is based upon the external and internal features of mouse embryos as described by Theiler,[18] Otis and Brent,[19] and unpublished observations by one of us (B.H.J.J.).

The embryos described by Theiler[18] were obtained by crossing an inbred strain of C57BL females, aged 3 to 5 months, with CBA males. Usually the first litters were utilized, but a few second litters were included. Those described by Otis and Brent[19] were offspring from a Bagg albino strain of mice derived from crosses of C57 females with DBA males. In both studies, copulation was determined by the presence of a vaginal plug which marked day one of gestation. Total gestation time in both series was a little over 19 days. It is to be noted that the total gestational time may vary slightly from one strain of mouse to another. In both series the embryos were measured before fixation, and Theiler[18] indicates that fixation in formalin may reduce the length by one-third. Carnegie stages 1 to 8 mouse embryos are summarized in Table 6.

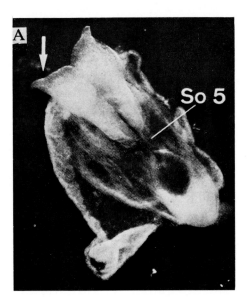

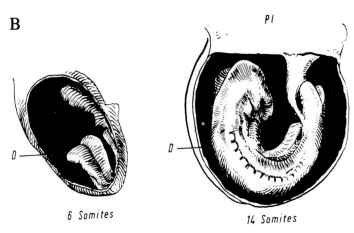

FIGURE 54   (A) Dorsal view of a 7-somite mouse embryo, i.e., an early Carnegie stage 10. Arrow = optic sulcus; So5 = somite 5. (B) Turning of the embryo. (Left drawing) before rotation of a 6-somite embryo (early Carnegie stage 10); (right drawing) after rotation of a 14-somite embryo (early Carnegie stage 11). Amnion is not shown. D = yolk sac (cut); PL = placenta. (From Theiler, K., *The House Mouse*, Springer-Verlag, New York, 1972. With permission.)

Mouse. Carnegie Stages 9 and 10 (Theiler stages 12 and 13)

Carnegie stage 9 embryos have an open neural plate and one to three pairs of somites. Theiler stage 12 embryos have a postcopulation (pc) age of about 9 days and have an open neural tube and 1 to 7 pairs of somites. Hence the first half of Theiler stage 12 is the equivalent of Carnegie stage 9.

Carnegie stage 10 embryos have 4 to 12 pairs of somites, beginning fusion of the neural folds in the future cervical region and otic placodes. Hence the second half of Theiler stage 12 and Theiler stage 13 are equivalent to Carnegie stage 10. The average pc age of these embryos is 9.5 days. Mouse embryos of Carnegie stage 9 and early stage 10 have a well marked "lordosis" which becomes converted into a ventral concavity as the embryo undergoes rotation. Viewed from the cranial towards the caudal end, the rotation proceeds clockwise along the body axis. Such a rotation does not occur in human embryos.

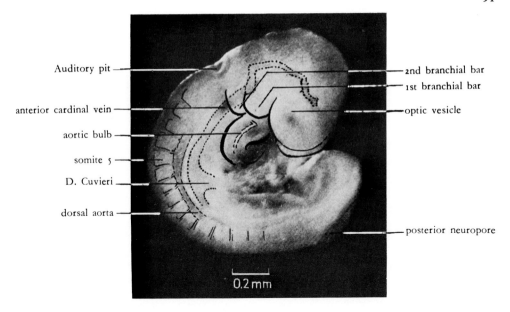

Auditory pit

2nd branchial bar

1st branchial bar

anterior cardinal vein

optic vesicle

aortic bulb

somite 5

D. Cuvieri

dorsal aorta

posterior neuropore

0.2 mm

FIGURE 55. Right aspect of a Carnegie stage 11 mouse embryo. Fourteen pairs of somites; 10 days pc. (From Theiler, K., *The House Mouse*, Springer-Verlag, New York, 1972. With permission.)

Mouse. Carnegie Stage 11 (Theiler stage 14)

Embryos of stage 11 vary in length from 1.2 to 2.5 mm and have a pc age of 10 days. They have 13 to 20 pairs of somites. Turning of the embryo is now complete. Mouse embryos of this stage have a more marked ventral curvature than human embryos. Also, they have a spiral form with the caudal end lying on the right side of the head, and this is the cause of the considerable variation in length. The neural tube is the main determinant of embryonic form. The cranial neuropore closes during this stage. The olfactory and lens placodes are appearing, and the otic placode is invaginating. Branchial bars one and two are present. An indistinct ridge, appearing at 15 somites, marks the beginning of the cranial limb bud, i.e., one stage earlier than in the human embryos. A long narrow vitelline duct is present.

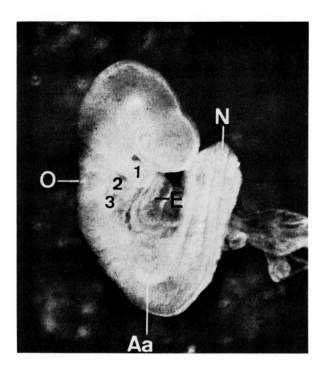

FIGURE 56.   Right aspect of a Carnegie stage 12 mouse embryo with 24 pairs of somites. O = otic vesicle separating from epidermis; Aa = cranial limb bud; N = suture line of neural fold; E = endocardium within bulbus arteriosus; 1, 2, 3 = branchial bars 1, 2, 3. (From Theiler, K., *The House Mouse,* Springer-Verlag, New York, 1972. With permission.)

Mouse. Carnegie Stage 12 (Theiler stage 15)

Embryos of stage 12 vary in length from 1.8 to 3.3 mm and have a pc age of 10.5 days. They have 21 to 29 pairs of somites. This stage is characterized by the presence of a distinct cranial limb bud opposite somites 8 to 12 and absence of a hindlimb bud. The third branchial bar is present. The otic vesicle is closed in the older members. The cranial neuropore is closed, but the caudal neuropore is still open. The lens placode and olfactory placode appear as distinct thickenings of the surface epithelium.

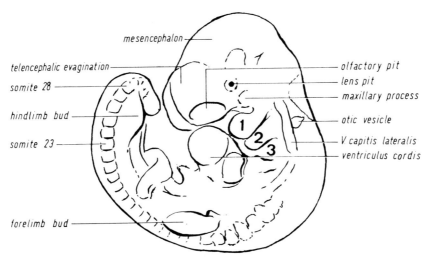

FIGURE 57.   Left aspect of a Carnegie stage 13 mouse embryo with 33 pairs of somites. (From Theiler, K., *The House Mouse,* Springer-Verlag, New York, 1972. With permission.)

Mouse. Carnegie Stage 13 (Theiler stage 16)

Embryos of stage 13 vary in length from 3.1 to 3.9 mm and have a pc age of 11 days. They have 30 to 34 pairs of somites. The caudal limb bud is now visible as a distinct bulge opposite somites 23 to 28. The beginning tail forms a short stump. The cervical sinus is appearing. The lens pit is beginning in the older specimens. The olfactory placode is beginning to indent. The otic vesicle is always closed. The caudal neuropore is closed.

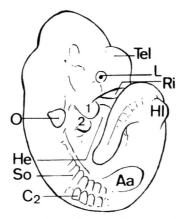

FIGURE 58.   Right aspect of a Carnegie stage 14 mouse embryo. Tel = cerebral hemisphere; L = lens invagination; Ri = olfactory pit; Hl = caudal limb bud; Aa = cranial limb bud; O = otic vesicle; He = heart; So = somites; C2 = ganglion cervicale 2; 1, 2 = branchial bars 1, 2. (From Theiler, K., *The House Mouse,* Springer-Verlag, New York, 1972. With permission.)

Mouse. Carnegie Stage 14 (Theiler stage 17)

Embryos of stage 14 vary in length from 3.5 to 4.9 mm and have a pc age of 11.5 days. They have 35 to 39 pairs of somites. The elongated limb buds curve forward and medially. The tail is considerably larger than that of its human counterpart. There is a distinct marginal lip around the olfactory placode. In the younger embryos the lens vesicles form deep pockets but have pore-like openings in the older ones. The first branchial bar is divided into maxillary and mandibular processes. The cranial somites are becoming indistinct. The cerebral vesicles are now distinctly seen.

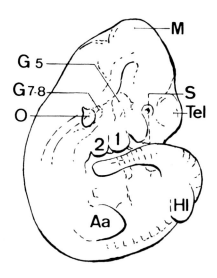

FIGURE 59.   Right aspect of a Carnegie stage 15 mouse embryo. M = mesencephalon; S = somites; Tel = cerebral hemisphere; G5 = trigeminal ganglion; G7-8 = ganglia of 7th and 8th cranial nerves; O = otic vesicle; Aa = cranial limb bud; Hl = caudal limb bud; 1 = mandibular process; 2 = hyoid bar. (From Theiler, I., *The House Mouse,* Springer-Verlag, New York, 1972. With permission.)

Mouse. Carnegie Stage 15 (Theiler stage 18)

Embryos of stage 15 vary from 5.0 to 6.0 mm in length and have an overage pc age of 12 days. Theiler[18] considers that his stage 18 mouse embryos can be placed into Carnegie stages 14 to 15, since Carnegie stage 15 human embryos have closed lens vesicles whereas closure of the mouse lens vesicles occurs during Theiler stage 18 of development. However, considering such internal criteria as separation of the trachea from the esophagus, elongation of the endolymphatic duct, the presence of a ureteric diverticulum, endocardial cushions, bulbar ridges, and a ventral pancreatic diverticulum, Theiler stage 18 embryos are equivalent to Carnegie stage 15. Carnegie stage 15 mouse embryos are characterized by the presence of distinct cerebral vesicles, a pontine flexure, a distinctly elongated endolymphatic duct, and a deepened olfactory pit.

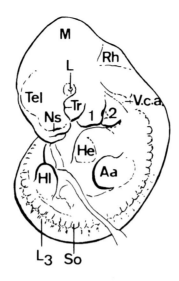

FIGURE 60. Left aspect of a Carnegie stage 16 mouse embryo. Tel = cerebral hemisphere; M = mesencephalon; Ns = nostril; Rh = rhombencephalon; L = lens vesicle (just closed); Tr = nasolacrimal groove; 1,2 = branchial bars; V.c.a. = vena cardinalis anterior; He = heart (ventricle); So = 1st lumbar somite; Aa = cranial limb bud; Hl = caudal limb bud; L_3 = 3rd lumbar spinal ganglion. (From Theiler, K., *The House Mouse*, Springer-Verlag, New York, 1972. With permission.)

Mouse. Carnegie Stage 16 (Theiler stage 19)

Based upon internal criteria such as the extent of heart, lung, kidney, and central nervous system development, embryos of Theiler stage 19 can be considered to be at Carnegie stage 16 of development. These embryos vary from 6.0 to 7.0 mm in length and have an average pc age of 12.5 days. They are characterized externally by the presence of a definite constriction dividing the cranial limb bud into a distal rounded footplate and a proximal region. At this stage the caudal limb bud is not yet distinctly subdivided. The olfactory pits are deep, resulting in the nostrils being narrowed to slits. The six auricular hillocks can now be distinctly seen. The caudal somites are still sharply defined, and the tail is considerably longer than in the previous stage. Retinal pigment is appearing.

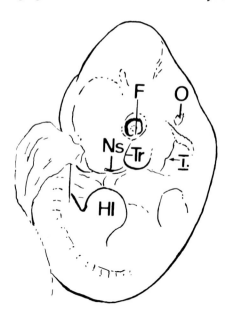

FIGURE 61.   Left aspect of a Carnegie stage 17/18 mouse embryo. O = otic vesicle; Ns = nostril; F = choroid fissure; 1 = first branchial pouch; Hl = caudal limb bud; Tr = nasolacrimal groove. (From Theiler, K., *The House Mouse,* Springer-Verlag, New York, 1972. With permission.)

Mouse. Carnegie Stages 17/18 (Theiler stage 20)

Each Theiler stage, up to stage 19, covers a half day of gestation, but subsequent stages until birth cover about one day of gestation. Examination of various internal features of Theiler stage 20 embryos such as the development of the central nervous system, inner ear, heart, metanephros, Meckel's cartilage, and clavicle indicates that Theiler stage 20 embraces both Carnegie stages 17 and 18.

Carnegie stage 17 embryos have an average pc age of 13 days and range in length from 7.0 to 8.0 mm. The forelimb footplate has toe rays, but its margin is not crenated. The hindlimb footplate is now distinctly demarcated. The lumbosacral somites are distinct but not the thoracic ones.

Carnegie stage 18 embryos have an average pc age of 13.5 days and range in length from 8.0 to 9.0 mm. They are distinguishable from the previous stage by having more pronounced toe rays and beginning crenation of the margin of the cranial footplate. There is only a slight indication of toe rays in the caudal footplate. The lumbosacral somites are now indistinct.

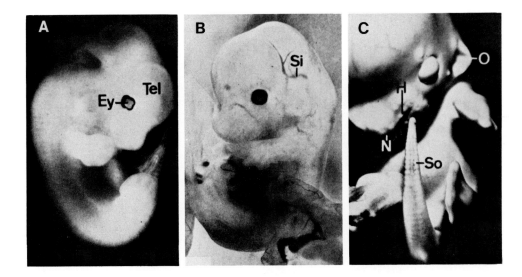

FIGURE 62. (A) Right view of a Carnegie stage 19 mouse embryo. Ey = eye; Tel = cerebral hemisphere. (B) Left view of a Carnegie stage 20 mouse embryo. Si = sigmoid sinus. (C) Frontal view of a Carnegie stage 20 mouse embryo. O = pinna; So = tail somites; N = nostril; H = rudiments of hair follicles (whiskers). (From Theiler, K., *The House Mouse*, Springer-Verlag, New York, 1972. With permission.)

Mouse. Carnegie Stages 19 and 20 (Theiler stage 21)

Theiler stage 21 embryos show the following features: fusion of the auricular hillocks to form a definitive pinna; the four upper rows of whiskers form conspicuous bumps, and there is a prominent hair follicle above each eye and another in front of each ear; the cartilaginous shaft of the humerus is at phase 3; the bucconasal membrane is ruptured; the sex of the gonad is recognizable histologically; the lens is solid; and, in older embryos, there are S-shaped metanephric vesicles. Hence, younger embryos of Theiler stage 21 are in Carnegie stage 19, and older ones in Carnegie stage 20.

Carnegie stage 19 embryos have a mean pc age of 14 days and range in length from 9.0 to 10.0 mm and can be staged according to the following criteria: somites are only visible in the tail; the eyelids are small ectodermal folds which do not yet cover the eyeball; the auricular hillocks have fused forming the pinna; the forelimb footplate is distinctly crenated, and whereas the hindlimb footplate has distinct toe rays, its margin is not crenated.

Carnegie stage 20 mouse embryos have a mean pc age of 14.5 days and range in length from 10.0 to 11.0 mm. They can be distinguished from Carnegie stage 19 embryos by their larger eyelids covering the upper and lower portions of each eyeball, a primordial hair follicle above each eye and another in front of each ear, the beginning separation of the fingers, and distinct crenation of the hindlimb footplate.

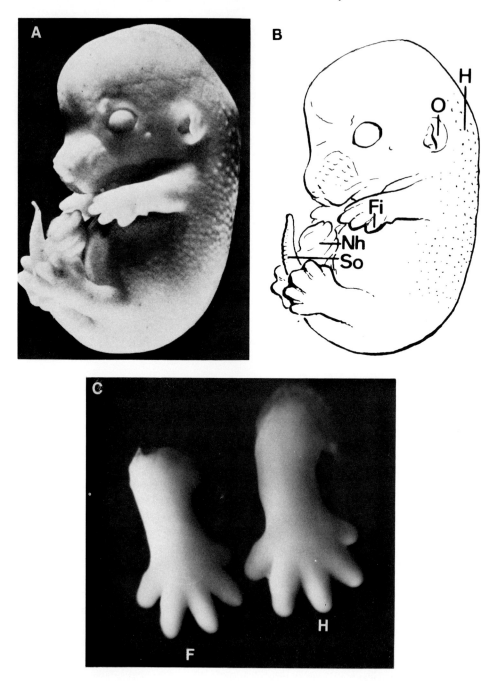

FIGURE 63.   (A) Left aspect of a Carnegie stage 21 mouse embryo. (From Theiler, I., *The House Mouse,*
Springer-Verlag, New York, 1972. With permission.) (B) Drawing of same embryo. O = pinna; H = hair
follicles; Fi = digit; Nh = umbilical hernia; So = tail somites. (From Theiler, K., *The House Mouse,* Springer-
Verlag, New York, 1972. With permission.) (C) Fore (F) and hind (H) limbs of a Carnegie stage 22 mouse
embryo.

Mouse. Carnegie Stage 21 and 22 (Theiler stage 22)

Theiler stage 22 embryos are characterized by non-fusion of the secondary palate and by the separation of the digits in the forelimb footplate and the beginning separation of the digits in the hindlimb footplate. This places them into Carnegie stages 21 and 22. Additionally, they show numerous young hair follicles in the skin except for the head region. The somites are only recognizable in the distal part of the tail. The growing pinna is turned forward and covers about one half of the external auditory meatus. The umbilical hernia is very conspicuous. Ossification has commenced in the humerus.

Carnegie stage 21 embryos have a mean pc age of 15 days and range in length from 10.5 to 11.5 mm. They have distinctly separated digits on the forelimb and deep indentation of the hindlimb footplate. The latter resembles the forelimb footplate of Carnegie stage 20 embryos.

Carnegie stage 22 embryos have a mean pc age of 15.5 days and range in length from 11.0 to 12.0 mm. The principal features distinguishing Carnegie stage 22 embryos from Carnegie stage 21 embryos is the separation of the distal phalanges of the toes of the hindfoot.

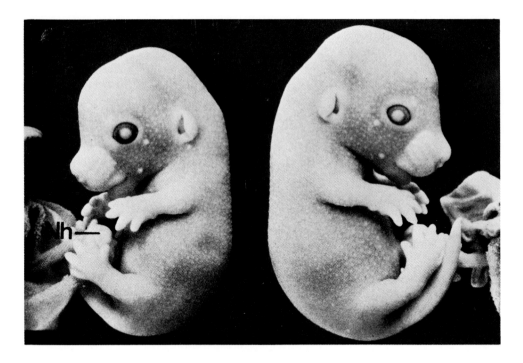

FIGURE 64.   Carnegie stage 23 mouse embryos. These are litter mates measuring 13.1 and 14.2 mm long, respectively. Nh = umbilical hernia. (From Theiler, K., *The House Mouse,* Springer-Verlag, New York, 1972. With permission.)

Mouse. Carnegie Stage 23 (Theiler stage 23)

Carnegie stage 23 embryos are characterized externally by complete separation of the toes in both limbs, distinct flexures at the elbow and knee, and well-defined eyelids that are about to close. Internal characteristics include a well-defined pia-arachnoid and an outburst of ossification including the scapula, humerus, radius, ulna, tibia, zygomatic, squamous temporal, palatine, frontal, and supra-occipital. The definitive feature of this stage is closure of the secondary palate.

All these features are present in Theiler stage 23 embryos, which have a mean pc age of 16 days and range in length from 12.0 to 14.0 mm. Additionally, the pinna covers more than one half of the external auditory meatus, and now hair follicles cover the entire body.

Carnegie Stages of Rat Embryos [Rattus rattus]

The development of the rat follows the same developmental schedule as the mouse, the major difference being that the rat conceptus implants a day and a half later than that of the mouse, resulting in a longer gestation period by one and a half days. Because of the similarity of development of the mouse and rat, the Carnegie stages of the rat embryo are summarized in Table 7. For details of Carnegie stage 9 to 23 rat embryos, refer to the description of these stages in the mouse. The data in Table 7 are from Witschi[20] and personal observations of one of us (B.H.J.J.). The morning on which the vaginal plug is found is considered day one of gestation.

Table 7
CARNEGIE STAGES 1 TO 23 OF
RAT EMBRYOS

Carnegie stage	Age (days)	Approximate length (mm)
1	1—2	—
2	3—4	—
3	5—6	—
4	7	—
5	8	—
6	9	—
7	9.5	—
8	10	1.0
9	10.5	1.5
10	11	2.0
11	11.5	2.5
12	12	3.5
13	12.5	4.5
14	13	5.5
15	13.5	6.0
16	14	6.5
17	14.5	8.5
18	15	9.5
19	15.5	10.5
20	16	12.0
21	16.5	14.0
22	17	15.0
23	17.5	16.0

Based on Witschi[20] and personal observations (B.H.J.J.).

VII. CHINESE HAMSTER AND GOLDEN HAMSTER

Carnegie Stages of Chinese Hamster Embryos [*Cricetulus griseus*]

In 1979 Donkelaar et al.,[21] described the developmental stages of the Chinese hamster embryo from the first appearance of the somites to the end of the embryonic period as indicated by closure of the secondary palate. The period covered corresponds to Carnegie stages 9 to 23, but they divided this period of hamster development into stages 9 to 19. Pickworth et al.[22] described the first 5 to 6 days of development, i.e., from fertilization to beginning implantation. No information is available for the period between early implantation and the first appearance of somites.

Up to stage 16, the Donkelaar stages match up with the Carnegie stages. However, Donkelaar et al.,[21] have only 3 stages of Chinese hamster development corresponding to the last seven Carnegie stages. In man, development from Carnegie stage 9 to 23 takes 41 days as opposed to 7 days in the Chinese hamster. Some features of the Chinese hamster develop a little earlier or a little later than in primates. However, these differences are no more than about one stage, which is less than can occur between littermates. Chinese hamster embryos are considerably shorter than human embryos. At stage 9 human embryos are 1.5 to 2.5 mm long, whereas Chinese hamster embryos are 0.7 mm, and at stage 23 their respective lengths are 27.0 to 31.0 mm and 12.0 to 14.0 mm.

Females were exposed to males for 3 hr, thus reducing the inexactitude of the postcopulation (pc) age to plus or minus 1½ hr. The age of the embryos was determined according to the number of days after copulation, and the age on the day of mating was recorded as embryonic day zero.

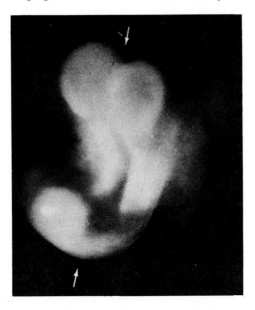

FIGURE 65.   Ventral view of a Carnegie stage 9 Chinese hamster embryo. Greatest length measured along the axis indicated by the arrows. Upper arrow is pointing at the cranial end of the neural groove. (From Donkelaar, H. J. ten, Geijsberts, L. G. M., and Dederen, P. J. W., *Anat. Embryol.*, 156, 1, 1979. With permission.)

Chinese Hamster. Carnegie Stage 9 (Donkelaar Stage 9)

A stage 9 Chinese hamster embryo is 0.7 mm in greatest length and has a pc age of 10 days. It has well-developed neural folds, and the first somites are present.

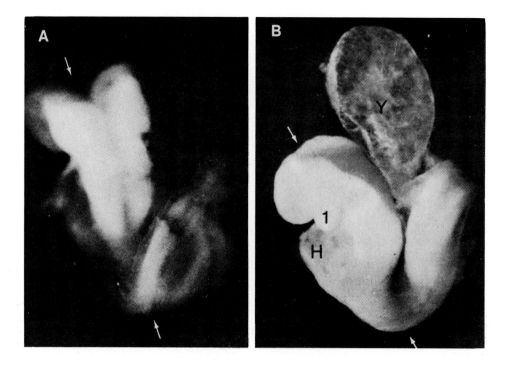

FIGURE 66. (A) Dorsal view of a Carnegie stage 10 Chinese hamster embryo. (B) Lateral view of a Carnegie stage 10 Chinese hamster embryo. H = heart anlage; Y = yolk sac; 1 = first branchial bar. Note the extreme "lordosis". Greatest length measured in the direction indicated by the arrows. (From Donkelaar, H. J. ten, Geijsberts, L. G. M., and Dederen, P. J. W., *Anat. Embryol.*, 156, 1, 1979. With permission.)

Chinese Hamster. Carnegie Stage 10 (Donkelaar Stage 10)

Embryos of stage 10 vary in greatest length from 1.2 to 2.0 mm and have a pc age of 10.5 to 11 days. They have 4 to 12 pairs of somites. The neural tube is beginning to fuse at the level of the future cervical region. The heart and yolk sac are prominent. There is a very marked "lordosis" which is converted into a "kyphosis" as a result of turning of the embryo. The first branchial bar is present. The otic placode is present and is invaginating in the older members of this group.

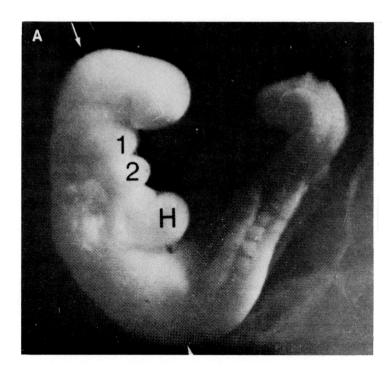

FIGURE 67.   (A) Lateral aspect of a Carnegie stage 11 Chinese hamster embryo.
1,2 = first and second branchial bars; H = heart anlage. Greatest length measured
in the direction of the arrows. (B) Same embryo cleared in methylbenzoate. * =
cranial flexure; arrow = otic invagination. (From Donkelaar, H. J. ten, Geijsberts,
L. G. M., and Dederen, P. J. W., *Anat. Embryol.,* 156, 1, 1979. With permission.)

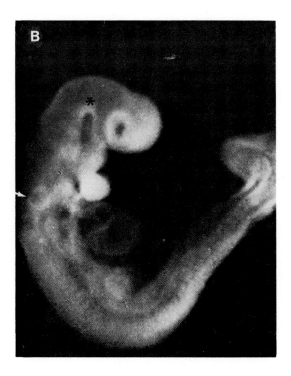

FIGURE 67B.

Chinese Hamster. Carnegie Stage 11 (Donkelaar Stage 11)

Embryos of stage 11 vary in greatest length from 2.0 to 2.5 mm in length and have a pc age of 10.75 to 11 days. They have 13 to 20 pairs of somites. The turning of the embryo is completed, and it now has a ventral concavity. The cranial neuropore is closed in the older members. Both mandibular and hyoid branchial bars are present. The heart is very prominent. The cranial limb bud is beginning to appear in the older specimens. The invaginated otic plate forms a deep open groove. The olfactory placode is beginning. There is a prominent cranial (or mesencephalic) flexure.

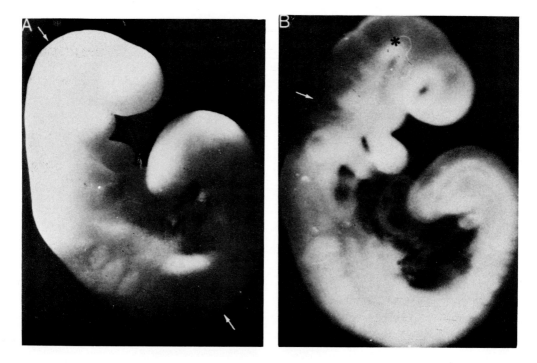

FIGURE 68.   (A) Lateral aspect of a Carnegie stage 12 Chinese hamster embryo. Greatest length measured in the direction of the arrows. (B) Lateral view of a Carnegie stage 12 Chinese hamster cleared in methylbenzoate. * = cranial flexure; arrow = beginning pontine flexure. (From Donkelaar, H. J. ten, Geijsberts, L. G. M., and Dederen, P. J. W., *Anat. Embryol.,* 156, 1, 1979. With permission.)

Chinese Hamster. Carnegie Stage 12 (Donkelaar Stage 12)

Embryos of stage 12 vary in greatest length from 2.7 to 3.2 mm and have a pc age of 11 to 11.5 days. They have 21 to 29 pairs of somites. All have a definite cranial limb bud, but only the older ones have the beginning of a caudal limb bud. Three branchial bars are present, and the maxillary process is developing. The body axis is now C-shaped following the turning of the embryo. The slit-like caudal neuropore is still open. The olfactory placode is more evident. The otic invagination is open in the younger members but is closed in the older ones. There is a well-developed cranial flexure, and the pontine flexure is just beginning.

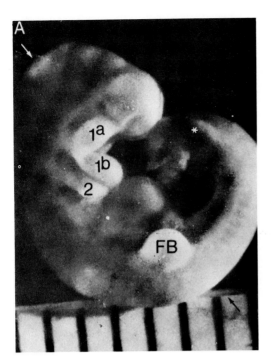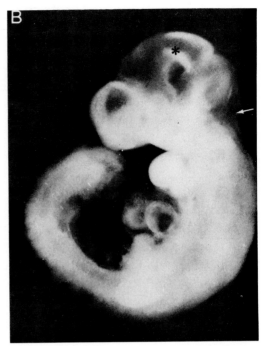

FIGURE 69. (A) Lateral view of a Carnegie stage 13 Chinese hamster embryo. 1a = maxillary process; 1b = mandibular process; 2 = second branchial bar; FB = forelimb bud; * = hindlimb bud. Greatest length measured in the direction of the arrows. (B) Lateral view of a Carnegie stage 13 Chinese hamster embryo cleared in methylbenzoate. * = cranial flexure; arrow = pontine flexure. (From Donkelaar, H. J. ten, Geijsberts, L. G. M., and Dederen, P. J. W., *Anat. Embryol.*, 156, 1, 1979. With permission.)

Chinese Hamster. Carnegie Stage 13 (Donkelaar Stage 13)

Embryos of stage 13 vary in greatest length from 2.8 to 4.2 mm and have a pc age of 11.5 to 12 days. The number of somites can no longer be accurately counted. The caudal neuropore is usually closed. The cranial limb bud is rapidly enlarging, and the caudal limb bud is clearly recognizable. The maxillary and mandibular processes are enlarging. The nasal pit is bounded by lateral and medial nasal processes. The lens placode is beginning to indent. The otic vesicle is always closed and completely detached from the overlying ectoderm. The cerebral hemispheres are beginning to evaginate. The cranial flexure is still the dominant one, but the pontine flexure is more marked. The older members have the beginnings of a cervical flexure.

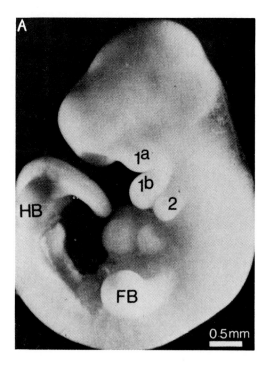 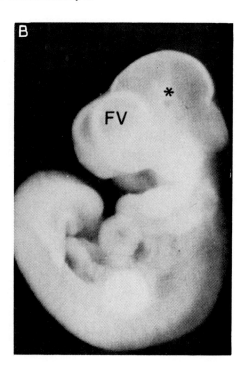

FIGURE 70.   (A) Lateral view of a Carnegie stage 14 Chinese hamster embryo. 1a = maxillary process; 1b = mandibular process; 2 = hyoid branchial bar; FB = cranial limb bud; HB = caudal limb bud. (B) Lateral view of a Carnegie stage 14 Chinese hamster embryo cleared in methylbenzoate. * = cranial flexure; FV = forebrain vesicle. (From Donkelaar, H. J. ten, Geijsberts, L. G. M., and Dederen, P. J. W., *Anat. Embryol.*, 156, 1, 1979. With permission.)

Chinese Hamster. Carnegie Stage 14 (Donkelaar Stage 14)

Carnegie stage 14 embryos vary in greatest length from 4.6 to 5.1 mm and have a pc age of 12 to 12.5 days. The limb and tail buds are rapidly enlarging, and the tail is considerably longer than that of human Carnegie stage 14 embryos. The cranial limb bud is divided into two parts, and there is a distinct caudal limb bud. The branchial bars are further enlarged. The deepening nasal pits are still situated laterally. A prominent feature is the distinct cervical flexure and a more distinct pontine flexure. The cerebral hemispheres are considerably enlarged. There is a deep lens indentation with a pore-like opening in the older members.

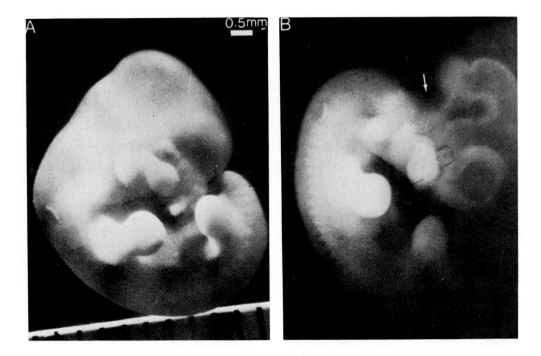

FIGURE 71. (A) Lateral view of a Carnegie stage 15 Chinese hamster embryo. (B) Lateral view of a Carnegie stage 15 Chinese hamster embryo cleared in methylbenzoate. Arrow = very deep pontine flexure. (From Donkelaar, H. J. ten, Geijsberts, L. G. M., and Dederen, P. J. W., *Anat. Embryol.,* 156, 1, 1979. With permission.)

Chinese Hamster. Carnegie Stage 15 (Donkelaar Stage 15)

Embryos of stage 15 vary in crown rump length from 5.1 to 6.0 mm and have a pc age of 12.5 to 13 days. The form of the embryo is dominated by the cervical flexure. The forelimb plate is beginning to flatten. The hindlimb is now divided into two parts and has elongated. There is a distinct tail. The nasal pits are deeper and still relatively widely separated and are bordered by well-developed medial and lateral nasal processes. The lens vesicle is closed. Retinal pigment can be seen in specimens cleared in methylbenzoate. The olfactory placode is considerably deepened. The forebrain vesicles are large, and the pontine flexure is very deep.

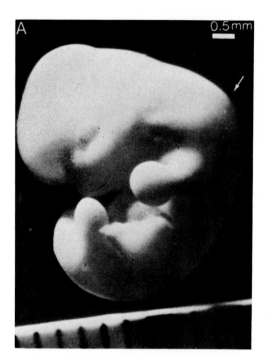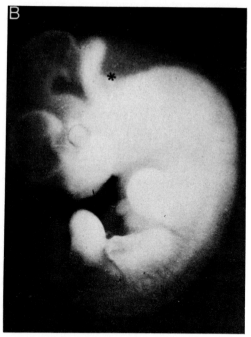

FIGURE 72.   (A) Lateral aspect of a Carnegie stage 16 Chinese hamster embryo. Arrow = cervical flexure. (B) Lateral view of a Carnegie stage 16 Chinese hamster embryo cleared in methylbenzoate. * = the deep pontine flexure. (From Donkelaar, H. J. ten, Geijsberts, L. G. M., and Dederen, P. J. W., *Anat. Embryol.*, 156, 1, 1979. With permission.)

Chinese Hamster. Carnegie Stage 16 (Donkelaar Stage 16)

Embryos of stage 16 vary in crown rump length from 5.6 to 6.3 mm and have a pc age of 13.5 days. The cervical flexure is still a dominant feature. The forelimb has a distinct footplate, and the hindlimb is flattening. There are now auricular hillocks on the hyoid bar. The nostrils are narrow slits, and the nasolacrimal groove is present. Retinal pigment is more obvious.

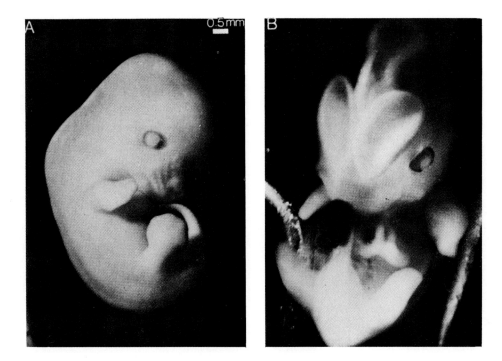

FIGURE 73. (A) Lateral view of a Carnegie stage 17/18 Chinese hamster embryo. Note the developing vibrissae. (B) Frontal view of same embryo cleared in methylbenzoate. Note the large telencephalic vesicles with their thick basolateral parts. (From Donkelaar, H. J. ten, Geijsberts, L. G. M., and Dederen, P. J. W., *Anat. Embryol.*, 156, 1, 1979. With permission.)

Chinese Hamster. Carnegie Stage 17/18 (Donkelaar Stage 17)

Embryos of stage 17/18 vary in length from 7.3 to 8.6 mm and have a pc age of 14 days. The paddle-shaped forelimb footplate now has digital rays. The hindlimb footplate is clearly demarcated from the distal part of the leg. The trunk is now straighter, and the head is moving up and decreasing the cervical flexure. Three to five rows of bumps indicate the commencement of the vibrissae. Retinal pigment is now more obvious, but no definite eyelids are present. The external auditory meatus is visible, but the prominent auricular hillocks are beginning to fuse. The nostrils are nearer the median plane. The umbilical hernia is beginning.

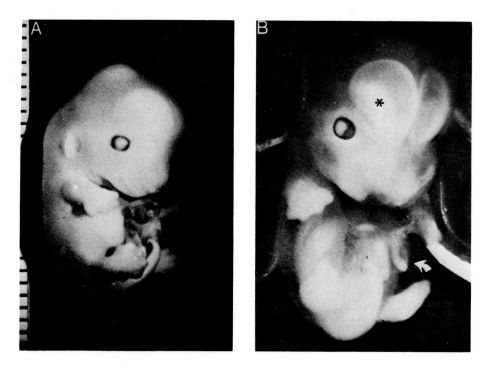

FIGURE 74.   (A) Lateral view of a Carnegie stage 19/20 Chinese hamster embryo. Bars on scale are 0.5 mm apart. (B) Frontal view of Carnegie stage 19/20 Chinese hamster embryo cleared in methylbenzoate. Arrow = umbilical hernia; * = thick basolateral part of the telencephalic vesicle. (From Donkelaar, H. J. ten, Geijsberts, L. G. M., and Dederen, P. J. W., *Anat. Embryol.*, 156, 1, 1979. With permission.)

Chinese Hamster. Carnegie Stage 19/20 (Donkelaar Stage 18)

Embryos of stage 19/20 vary in crown rump length from 8.5 to 10.0 mm and have a pc age of 15 to 15.25 days. The forelimb footplate is clearly indented, but the hindlimb footplate is just becoming indented. The trunk has lengthened and straightens slightly as the head is lifted. A definitive auditory pinna is present, and the external auditory meatus is clearly visible. The eyelids are beginning to form. The nostrils are close to the midline. The lower jaw is still relatively small. The umbilical hernia is larger.

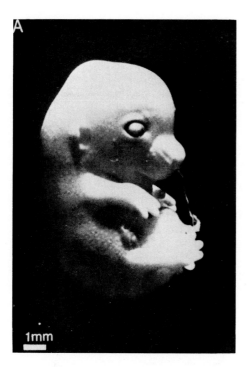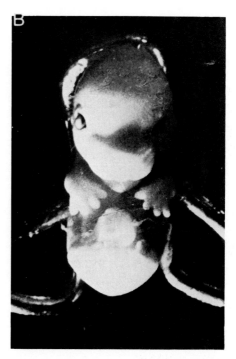

FIGURE 75. (A) Lateral view of a Carnegie stage 22 Chinese hamster embryo. (B) Frontal view of same embryo. Note the large umbilical hernia. (From Donkelaar, H. J. ten, Geijsberts, L. G. M., and Dederen, P. J. W., *Anat. Embryol.*, 156, 1, 1979. With permission.)

Chinese Hamster. Carnegie Stages 21 to 23 (Donkelaar Stage 19 and Fetal Stage 1)

Embryos of stages 21 to 22 vary in crown rump length from 10.5 to 11.0 mm and have a pc age of 15.75 to 16 days. The individual digits are now clearly separated in the forelimb but not in the hindlimb. There are numerous hair follicles on the trunk but only a few on the head. The eyelids, particularly the upper ones, are very prominent. The enlarged pinna is turned forward and covers half of the external auditory meatus. The lower jaw has increased markedly in size. There is still a prominent umbilical hernia. The secondary palate is fusing. The eyelids cover only about one fifth of the exposed surface of the eye, and complete closure does not occur until day 18.

Embryos of Carnegie stage 23 (Donkelaar's fetal stage 1) vary in crown rump length from 12.0 to 14.0 mm and have a pc age of 17 days. The toes are separated in both limbs, and the degree of eye development corresponds to that of Carnegie Stage 23 in primates. Also, it closely corresponds with mouse stage 23.

Table 8
CHRONOLOGY OF THE CARNEGIE STAGES OF CHINESE AND GOLDEN HAMSTER EMBRYOS

Chinese hamster age (days)	Carnegie stage	Golden hamster age (days)	Figure numbers in Boyer[23]
10	9	7.75	5.1.1
10.5	10	8—8.25	5.1.2
10.75	11	8.5	5.1.3
11—11.5	12	8.75	5.1.4
11.5—12	13	9	5.1.5
12—12.5	14	9.25—9.5	5.1.6
12.5—13	15	9.5—10	5.1.7
13.5	16	10.5	5.1.8, 5.2.9
14	17	11—11.5	5.2.10
	18		5.2.11
15	19	11.5—12	5.2.12
16	21	12—12.5	5.2.13
	22		5.2.14
17	23	13	5.2.15

Based on Donkelaar et al.[21] and Boyer.[23]

Carnegie Stages of Golden Hamster Embryos *[Mesocricetus Auratus]*

Boyer[23] described golden hamster embryos collected at 6-hr intervals from 7.5 days (the last presomite stage) to 10 days pc and at 12-hr intervals thereafter. From the first somite stage to closure of the secondary palate he described 16 stages, of which fifteen were illustrated. Type specimens for each chronological age were selected by inspection of external features, shape, visceral bars, etc., and by measurement of greatest overall length. However, he does not record these lengths but has a 2.0 mm scale line alongside the photograph of each embryo. From these scales it is estimated that the Carnegie stage 9 embryo is 2.0 mm long, and the stage 23 embryo is about 16.0 mm. Thus, there is probably little size difference between mouse, Chinese hamster, and golden hamster.

The external and internal features of the various Carnegie stages of the Chinese and golden hamsters are so much alike that it was deemed unnecessary to illustrate the stages of the golden hamster. The major difference, however lies in the ages of the various stages, since the time between stages 9 and 23 is seven days in the Chinese hamster but only five days in the golden hamster. Table 8 tabulates these differences and lists Boyer's[23] figures of these various stages.

Table 9
CARNEGIE STAGES 1 TO 8 OF
GUINEA PIG EMBRYOS

Carnegie stage	Age (days)	Features
1	1	Fertilization
2	3.5	From 2 to 8 cells
3	5	Free blastocyst
4	6	Implantation
5	6	Bilaminar embryonic disc
6	11.75	Primitive streak
7	12	Notochordal process
8	13.5	Neural plate

Based on Scott.[26]

VIII. GUINEA PIG

Carnegie Stages of Guinea Pig Embryos *[Cavia Porcellus]*

The papers of Harman and Prickett[24] and Harman and Dobrovolny[25] describe the development of the external form of the guinea pig embryo and fetus between 11 and 35 days of gestation. Scott[26] published a table of normal development of the guinea pig from fertilization to an embryo of some 26 days, based on his own work and that of several other investigators. He described both internal and external features, and this makes it possible to arrange eleven of his embryos into Carnegie stages 9 to 21. Harman and Dobrovolny[25] describe and illustrate a 29-day fetus which is, in fact, a Carnegie stage 23 embryo.

Females were exposed to males between 2000 and 2200 hr and were separated when copulation had taken place. Ages were given in days, hours, and minutes from the observed copulation. Embryos were taken at approximately 24-hr intervals from 10 to 30 days after copulation. The most advanced embryo from each litter was selected for examination. All the figures were redrawn from Scott.[26] The following descriptions are based upon only one embryo at each of the Carnegie stages described. The major features of embryos of Carnegie stages 1 to 8 are summarized in Table 9.

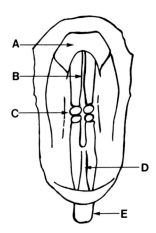

FIGURE 76.   Dorsal aspect of a Carnegie stage 9 guinea pig embryo. A = neural plate; B = notochordal canal; C = somite; D = primitive streak; E = allantois.

Guinea Pig. Carnegie Stages 9 and 10

A stage 9 embryo has a postcopulation (pc) age of about 14.5 days. The body is elongated and has an elevated brain plate and neural folds. Two pairs of somites are present. A stage 10 embryo with a pc age of 14.5 days has five pairs of somites and beginning closure of the neural tube.[27]

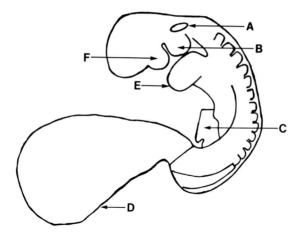

FIGURE 77. Left aspect of an early Carnegie stage 11 guinea pig embryo. A = otic vesicle; B = second branchial bar; C = yolk stalk; D = allantois; E = heart; F = first branchial bar.

Guinea Pig. Carnegie Stage 11

An early stage 11 embryo has a pc age of 15.5 days and is about 4.0 mm in length. Thirteen pairs of somites are present. The otic vesicle is widely open. The first and second branchial bars are present, and the cervical sinus is widely open. The caudal neuropore is still open. The pear-shaped allantois is almost as large as the embryo proper.

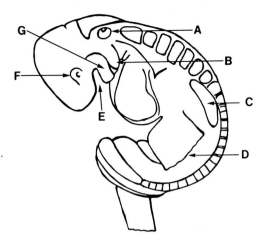

FIGURE 78.   Left aspect of an early Carnegie stage 12 guinea pig embryo. A = otic vesicle; B = second branchial bar; C = cranial limb bud; D = yolk stalk; E = mouth; F = eye; G = mandibular process.

Guinea Pig. Carnegie Stage 12 (Early)

A stage 12 embryo has a pc age of 16.5 days and is about 5.4 mm in length. Twenty three pairs of somites are present. There is a well-developed yolk sac. The first branchial bar is beginning to divide into dorsal and ventral parts. The slightly sunken cervical sinus contains the third branchial bar and a rudiment of the fourth. The otic vesicle is almost closed off. The cranial limb bud presents as a narrow thickening at the base of somites five to ten. The olfactory placode is just appearing. The caudal neuropore is closed.

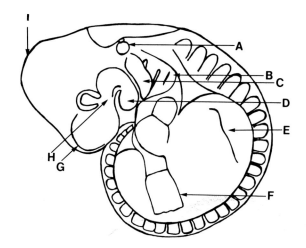

FIGURE 79. Left aspect of a late Carnegie stage 12 guinea pig embryo. A = otic vesicle, B = fourth branchial bar; C = second branchial bar; D = mandibular process; E = cranial limb bud; F = yolk stalk; G = telencephalon; H = maxillary process; I = mesencephalon.

Guinea Pig. Carnegie Stage 12 (Late)

A late stage 12 embryo has a pc age of 17.5 days and is about 6.0 mm in length. Twenty-nine pairs of somites are present, and given one more somite it would be in stage 13. There is a slight cervical flexure, the large, rounded mandibular process touches the cardiac bulge, and the smaller maxillary process touches the eye. The second branchial bar is now divided into dorsal and ventral parts. The fourth branchial bar is seen within the cervical sinus. There is a thick nasal placode. The tail is thick and has a terminal knob. The cranial limb bud is rounded in section, and an indefinite thickening marks the beginning of the caudal limb bud.

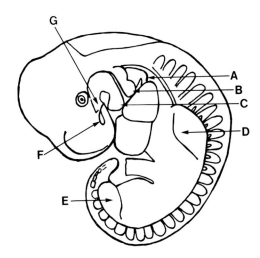

FIGURE 80.   Left aspect of a Carnegie stage 13 guinea pig embryo. A = fourth branchial bar; B = second branchial bar; C = mandibular process; D = cranial limb bud; E = caudal limb bud; F = nasal pit; G = maxillary process.

Guinea Pig. Carnegie Stage 13

A stage 13 embryo has a pc age of 18.5 days and is just over 6.0 mm in length. Thirty-one pairs of somites are present. The cranial limb bud has lengthened, and the caudal limb bud is now well defined. The cervical and sacral flexures are present. The olfactory placode is beginning to invaginate.

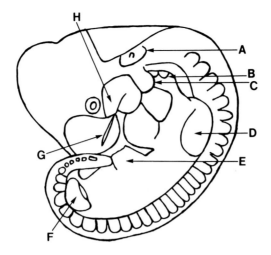

FIGURE 81.   Left aspect of a Carnegie early stage 14 guinea pig embryo. A = otic vesicle; B = fourth branchial bar; C = second branchial bar; D = cranial limb bud; E = yolk stalk; F = caudal limb bud; G = nasal pit; H = maxillary process.

Guinea Pig. Carnegie Stage 14 (Early)

A stage 14 embryo has a pc age of 19 days and is just under 7.0 mm in length. Thirty-five pairs of somites are present; hence it is just at the beginning of stage 14. The cranial limb bud has elongated in a ventral direction, and the caudal limb bud now has a distinct ventral edge. The maxillary process is now nearly as long as the mandibular process. There is a long deep nasal pit closed at its posterior end. The cervical sinus is deep. The lens pit is still open. The completely separated otic vesicle has a well-developed endolymphatic duct.

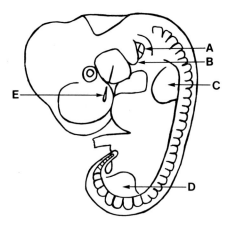

FIGURE 82.   Left aspect of a Carnegie late stage 14 guinea pig embryo. A = third branchial bar; B = second branchial bar; C = cranial limb bud; D = caudal limb bud; E = nasal pit.

Guinea Pig. Carnegie Stage 14 (Late)

A stage 15 embryo has a pc age of 19.7 days and is about 8.0 mm in length. Thirty-eight pairs of somites are present, so this embryo is on the verge of entering stage 15. The cervical flexure is now about 120°, and the tail is pointed. The lens vesicles are closed. The nasal pit is narrow. The cranial limb bud is flattened laterally near its tip, and the marginal vein is present. The caudal limb bud is rounded at its tip.

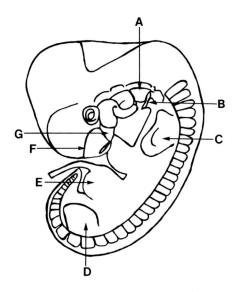

FIGURE 83.   Left aspect of a Carnegie stage 15 guinea pig embryo. A = second branchial bar; B = cervical sinus; C = cranial limb bud; D = caudal limb bud; E = umbilical hernia; F = premaxillary process; G = maxillary process.

Guinea Pig. Carnegie Stage 15

This stage 15 embryo has a pc age of 20.75 days and is about 12.0 mm in length. Thirty-nine pairs of somites are said to be visible,[26] which is fewer than expected for stage 15. However, the beginning delineation of the forelimb plate places it in stage 15. This is confirmed by the degree of development of a number of internal features, e.g., a ureteric diverticulum, bulbar ridges, etc. The cervical sinus is reduced to a slit, and the last visible branchial bar is the second.

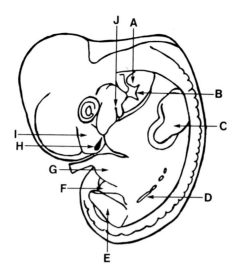

FIGURE 84.   Left aspect of a Carnegie stage 16/17 guinea pig embryo. A = pinna; B = external auditory meatus; C = cranial limb bud; D = mammary ridge; E = caudal limb bud; F = genital tubercule; G = umbilical hernia; H = nasal pit; I = premaxillary process; J = mouth.

Guinea Pig. Carnegie Stage 16/17

A stage 16/17 embryo has a pc age of 21.7 days and is just over 12.0 mm in length. The more cranial somites are indistinct. The cervical flexure is slightly less than 90°, and the head is roughly square. The mammary ridge is seen. The forelimb footplate is flattened, but the hindlimb footplate is not yet formed. Jacobsen's organs are present, thus indicating stage 17. The external auditory meatus is seen, and the auricular hillocks are beginning to fuse to form the pinna.

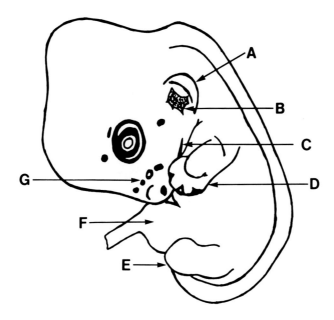

FIGURE 85. Left aspect of a Carnegie stage 19 guinea pig embryo. A = pinna; B = external auditory meatus; C = mouth; D = forelimb; E = hindlimb; F = umbilical hernia; G = vibrissae.

Guinea Pig. Carnegie Stage 19

A stage 19 embryo has a PC age of 23.7 days and is about 12.6 mm in length. Somites are no longer visible. There is a large umbilical hernia. The cervical flexure is reduced to 45°, and there is a prominent lumbar spine. The oronasal groove has disappeared and the eyelids are appearing. The cervical sinus has closed. A definitive pinna is present. Several papillae mark the beginnings of the vibrissae. Toe rays are present on the forefoot, but the hindlimb footplate is not yet crenated.

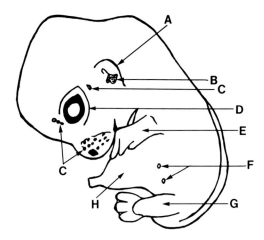

FIGURE 86.   Left aspect of a Carnegie stage 21 guinea pig embryo. A = pinna; B = external auditory meatus; C = vibrissae; D = eyelid; E = forelimb; F = nipples; G = hind-limb; H = umbilical hernia.

Guinea Pig. Carnegie Stage 21

A stage 21 embryo has a pc age of 26 days and is about 16.5 mm long. The cervical flexure is now less than 45°. The eyelids and snout are prominent. The edge of the pinna is rounded and curved over, partially covering the external auditory meatus. The vibrissae have increased in number. The toes are clearly demarcated on both fore and hind feet.

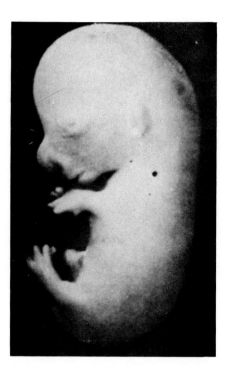

FIGURE 87. Left aspect of a Carnegie stage 23 guinea pig embryo (From Harman, M. T. and Dobrovolny, M., *J. Morphol.*, 54, 493, 1933. With permission.)

Guinea Pig. Carnegie Stage 23

Stage 23 embryos have a pc age of 29 days, and they average 20.9 mm in length. There is no longer a cephalic flexure. The tail is beginning to shorten. The eyelids are beginning to fuse. The pinna is elongating and bending forward to almost cover the external auditory meatus. The nose is long and somewhat beak-like. The external genitalia have assumed their specific characteristics.

Table 10
CARNEGIE STAGES 1 TO 8 OF
RABBIT EMBRYOS

Carnegie stage	Age (days)	Features
1	1	Fertilization
2	2	From 2 to about 16 cells
3	3	Free blastocyst
4	4	Attaching blastocyst
5	5	Bilaminar embryonic disc
6	6	Beginning primitive streak
7	6.5	Notochordal process
8	7.5	Neural plate

Based on Hartman.[28]

IX. RABBIT

Carnegie Stages of Rabbit Embryos *[Oryctolagus cuniculus]*

Very little has been written about the embryology of the rabbit, despite its widespread use in biomedical research.[28] The following description is based mainly on the work of Edwards.[29] supplemented by internal features of the rabbit embryo described by Marshall[30] and data from Hartman.[28]

Edwards[29] described the development of the rabbit between 7.7 to 21 days of gestation. She divided this period of development into 17 stages, of which stages 1 to 15 can be equated with Carnegie stages 9 to 23. The boundaries between her stages are generally somewhat different from those used for the Carnegie stages, so that in some instances two of her stages equate with one Carnegie stage. However, Carnegie stages 9 and 23 rabbit embryos can be definitely established, and on the basis of Edwards'[29] description of external features and Marshall's[30] description of some internal features, it is possible to approximately match up the intervening stages. The main features of Carnegie stage 1 to 8 rabbit embryos are summarized in Table 10.

Virgin does were mated with breeding bucks, and the exact time of copulation noted. Rabbits undergo copulatory-stimulated ovulation, and day zero of gestation is considered to begin at the time of copulation.

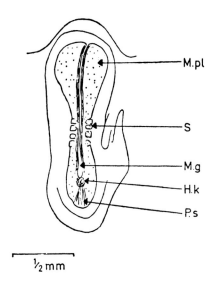

FIGURE 88.   Dorsal view of a Carnegie stage 9 rabbit embryo. Three pairs of somites are present and the fourth pair are just appearing. H.k = primitive node; M.g = neural groove; M.pl = neural plate; P.s = primitive streak; S = somite. (From Edwards, J. A., *Advances in Teratology,* Vol. 3, Woollam, D. H. M., Ed., Academic Press, New York, 1968, Chap. 7. With permission.)

Rabbit. Carnegie Stage 9 (Edwards Stage 1)

Embryos of stage nine have a postcopulation (pc) age of 7.7 to 8.25 days and have 1 to 4 pairs of somites. The neural plate is open throughout this stage and ends caudally at the primitive node, which is visible as a dense patch of tissue. The neural plate is flattened and somewhat expanded laterally in the head region.

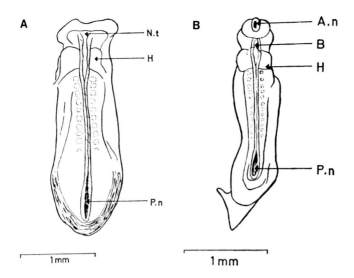

FIGURE 89.   (A) A Carnegie stage 10 rabbit embryo (Edwards stage 2). Nine pairs of somites are present, and the tenth pair is appearing. H = heart; N.t = neural tube; P.n = caudal neuropore. (B) A Carnegie stage 10 rabbit embryo (Edwards stage 3). Twelve pairs of somites are present. A.n = cranial neuropore; B = brain; H = heart; P.n = caudal neuropore. (From Edwards, J. A., *Advances in Teratology,* Vol. 3, Woollam, D. H. M., Ed., Academic Press, New York, 1968, Chap. 7. With permission.)

Rabbit. Carnegie Stage 10 (Edwards Stages 2 and 3)

Embryos of stage 10 have a greatest length of about 2.0 mm and a pc age of 8.25 to 9 days. Four to twelve pairs of somites are present. Fusion of the neural tube is beginning in the future cervical region. The otic placodes and optic vesicles appear in this stage.

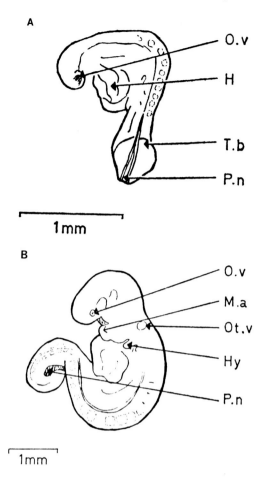

FIGURE 90.   (A) A Carnegie stage 11 rabbit embryo (Edwards stage 4). H = heart; O.v = optic vesicle; P.n = caudal neuropore; T.b = tail bud. (B) A Carnegie stage 11 rabbit embryo (Edwards stage 5). M.a = 1st branchial bar; Hy = 2nd branchial bar; O.v = optic vesicle; Ot.v = otic vesicle; P.n = caudal neuropore. (From Edwards, J. A., *Advances in Teratology,* Vol. 3, Woollam, D. H. M., Ed., Academic Press, New York, 1968, Chap. 7. With permission.)

Rabbit. Carnegie Stage 11 (Edwards Stages 4 and 5)

Embryos of stage 11 have a length of about 3.0 mm and a pc age of 9 to 9.75 days. They have 13 to 20 pairs of somites, and the first and second branchial bars are present. The cranial neuropore closes during this stage. The olfactory and lens placodes appear, and the otic placode begins to invaginate. Turning of the embryo commences at the beginning of this stage and is complete by its end, and then the embryo becomes C-shaped. The tail curves over to the right, and the caudal neuropore is still open.

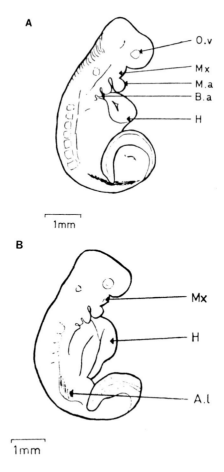

FIGURE 91. (A) A Carnegie stage 12 rabbit embryo (Edwards stage 6). Mx = maxillary process; M.a = mandibular process; B.a = 3rd branchial bar; H = heart; O.v = optic vesicle. (B) A Carnegie stage 12 rabbit embryo (Edwards stage 7). A.l = cranial limb bud; H = heart; Mx = maxillary process. (From Edwards, J. A., *Advances in Teratology,* Vol. 3, Woollam, D. H. M., Ed., Academic Press, New York, 1968, Chap. 7. With permission.)

Rabbit. Carnegie Stage 12 (Edwards Stages 6 and 7)

Embryos of stage 12 have a length of between 4.0 and 5.0 mm and a pc age of 9.75 to 11 days. They have 21 to 29 pairs of somites. First the 3rd and then the 4th branchial bars appear, and in the latter part of this stage, the cervical sinus appears. The cranial limb bud arises during the latter part of this stage. A distinct lens placode is present. Internal features include the beginnings of the thyroid gland, the lung bud, and the appearance of Rathke's pouch.

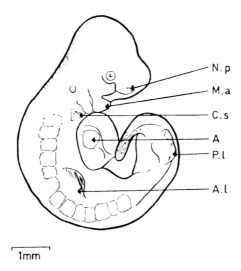

FIGURE 92.   Carnegie stage 13/14 rabbit embryos (Edwards stage 8). A = auricle; A.l = cranial limb bud; C.s = cervical sinus; M.a = mandibular process; N.p = nasal pit; P.l = caudal limb bud. (From Edwards, J. A., *Advances in Teratology,* Vol. 3, Woollam, D. H. M., Ed., Academic Press, New York, 1968, Chap. 7. With permission.)

Rabbit. Carnegie Stage 13/14 (Edwards Stage 8)

Embryos of stage 13/14 have a length of about 5.5 mm and a pc age of 11 to 12 days. The caudal limb buds appear at the beginning of this stage as small ridges on either side of the caudal end of the embryo. The nasal pits face laterally. The maxillary processes are prominent, and the mandibular processes approach the midline and begin to fuse. The third and fourth branchial bars are deep within the cervical sinus. The otic vesicle is closed.

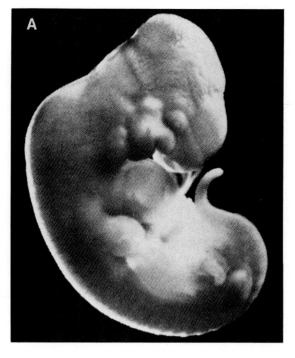

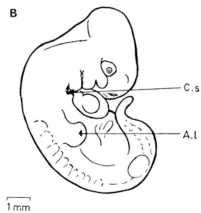

FIGURE 93. (A) Photograph of a Carnegie stage 15 rabbit embryo. (B) Diagram of same embryo. C.s = cervical sinus; A.l = cranial limb bud. (From Edwards, J. A., *Advances in Teratology,* Vol. 3, Woollam, D. H. M., Ed., Academic Press, New York, 1968, Chap. 7. With permission.)

Rabbit. Carnegie Stage 15 (Edwards Stage 9)

Stage 15 embryos have a length of about 8.0 mm and have a pc age of 12 to 13 days. The cranial limb buds have a constriction which marks off the footplate. At the beginning of this stage the maxillary process is prominent, and later it meets the lateral nasal process and the nasolacrimal groove is formed. The third and fourth branchial bars are now completely enclosed within the cervical sinus, which is reduced to a small aperture behind the second branchial bar. The lens vesicle is closing, and there is a well-developed pontine flexure.

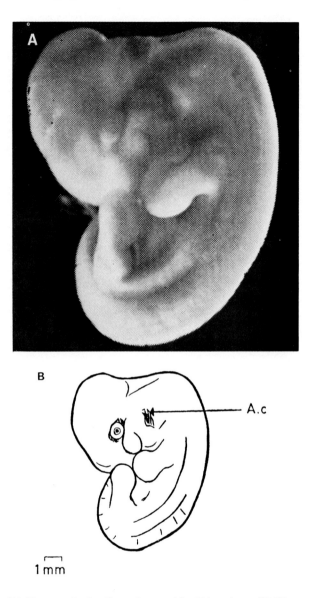

FIGURE 94.   (A) Photograph of a Carnegie stage 16 rabbit embryo. (B) Diagram of same embryo. A.c = auditory cleft. (From Edwards, J. A., *Advances in Teratology,* Vol. 3, Woollam, D. H. M., Ed., Academic Press, New York, 1968, Chap. 7. With permission.)

Rabbit. Carnegie Stage 16 (Edwards Stage 10)

Carnegie stage 16 embryos have a length of about 9.0 mm and a pc age of 13 to 14 days. The cranial limb bud has a distinctly rounded footplate, and the footplate of the caudal limb bud is beginning to be delineated. Auricular hillocks are appearing on the caudal side of the mandibular process and the cranial side of the hyoid bar. These two structures meet below the auditory cleft but have not yet fused. The head is tucked in, and the nasal region touches the heart. The lens vesicle is completely closed. The cervical somites are now indistinct. The milk ridges are now visible.

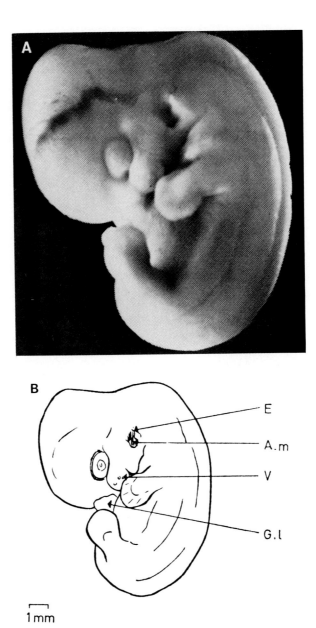

FIGURE 95. (A) Photograph of a Carnegie stage 17/18 rabbit embryo. (B) Diagram of same embryo. A.m = auditory meatus; E = auricle; G.l = umbilical hernia; V = vibrissae. (From Edwards, J. A., *Advances in Teratology*, Vol. 3, Woollam, D. H. M., Ed., Academic Press, New York, 1968, Chap. 7. With permission.)

Rabbit. Carnegie Stage 17/18 (Edwards Stage 11)

Carnegie stage 17/18 embryos have a length of about 10.5 mm and a pc age of 14 to 15 days. Retinal pigment is now visible, and the auricular hillocks are fusing to form the auricle, which appears as a slight ridge behind the external auditory meatus. The first vibrissae are seen on the maxillary process. The footplate of the cranial limb now has distinct toe rays, and they begin to appear in the caudal footplate at the end of this stage. There is a prominent umbilical hernia. Eyelids and two pairs of mammary buds appear in the older members of this stage.

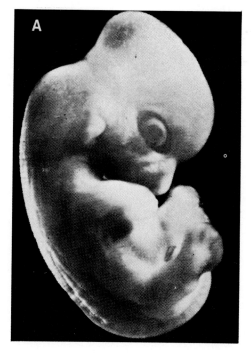

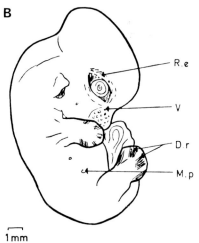

1 mm

FIGURE 96.   (A) Photograph of a Carnegie stage 19 rabbit embryo. (B) Diagram of same embryo. D.r = digital rays; M.p = mammary buds; R.e = upper eyelid; V = vibrissae. (From Edwards, J. A., *Advances in Teratology*, Vol. 3, Woollam, D. H. M., Ed., Academic Press, New York, 1968, Chap. 7. With permission.)

Rabbit. Carnegie Stage 19 (Edwards Stage 12)

Carnegie stage 19 embryos have a length of about 14.0 mm and a pc age of 15 to 16 days. They are characterized by the presence of definitive eyelids, a distinct projecting auricle, crenation of the cranial footplate, and distinct toe rays in the caudal footplate. There are four rows of vibrissae on the maxillary process, and the number in each row varies from two to five. The eyelids are more prominent, and a vibrissa develops above each eye. The enlarged auricle partially covers the external auditory meatus. The umbilical hernia increases in size, and a third pair of mammary buds appear. The back of the embryo is straightening, and the snout is elongating.

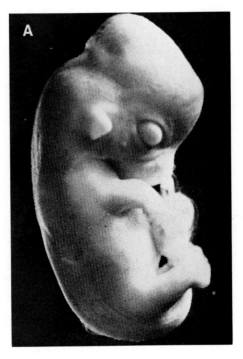

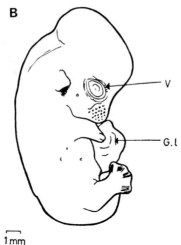

FIGURE 97. (A) Photograph of a Carnegie stage 20/21 embryo. (B) Diagram of same embryo. G.l = umbilical hernia; V = vibrissae. (From Edwards, J. A., *Advances in Teratology,* Vol. 3, Woollam, D. H. M., Ed., Academic Press, New York, 1968, Chap. 7. With permission.)

Rabbit. Carnegie Stage 20/21 (Edwards Stage 13)

Stage 20/21 embryos have a length of about 16.0 mm and a pc age of 16 to 17 days. There is beginning separation of the toes of the cranial footplate, and crenation of the caudal footplate starts at the early part of this stage. Separation of the toes of the caudal footplate occurs towards the end of this stage. The eyelids are beginning to cover the eyes. The auricle has increased in size but does not completely cover the external auditory meatus. There are two vibrissae over each eye, and the number of vibrissae on the maxillary process increases. A fourth pair of mammary buds appear. The umbilical hernia increases in size. There is further straightening of the back of the embryo and elongation of the snout.

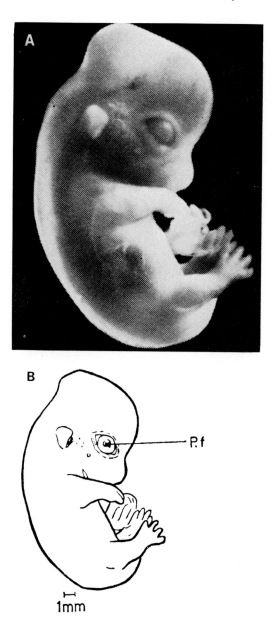

FIGURE 98.   (A) Photograph of a Carnegie stage 22 rabbit embryo. (B) Diagram of same embryo. P.f = palpebral fissure. (From Edwards, J. A., *Advances in Teratology*, Vol. 3, Woollam, D. H. M., Ed., Academic Press, New York, 1968, Chap. 7. With permission.)

Rabbit. Carnegie Stage 22 (Edwards Stage 14)

Stage 22 embryos have a length of about 22.0 mm and a pc age of 17 to 17.5 days. The toes of the forefoot are completely separated, but only the distal phalanges are separated in the hindfoot. The ankle and wrist are recognizable. The eyelids cover about one-third of the eyeball. The umbilical hernia is smaller. The number of vibrissae on the nasal region increases. There is now a row of four vibrissae over each eye.

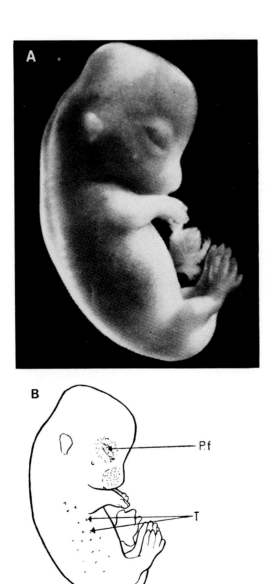

FIGURE 99. (A) Photograph of a Carnegie stage 23 rabbit embryo. (B) Diagram of same embryo. P.f = palpebral fissure; T = hair follicles on trunk. (From Edwards, J. A., *Advances in Teratology*, Vol. 3, Woollam, D. H. M., Ed., Academic Press, New York, 1968, Chap. 7. With permission.)

Rabbit. Carnegie Stage 23 (Edwards Stage 15)

Stage 23 embryos have a length of about 24.0 mm and a pc age of 17.5 to 18.75 days. The secondary palate has fused, and there is complete separation of the toes in both cranial and caudal feet. The eyelids cover most of the eye and are just about to fuse. Hair follicles are present all over the trunk but not on the head. The umbilical hernia is much reduced.

Table 11
CARNEGIE STAGES 1 TO 8 OF
SHEEP EMBRYOS

Carnegie stage	PC age (days)	Features
1	1	Fertilization
2	2—4	2 — 16 cells
3	5—6	Blastocyst
4	7—8	Endoderm formation
5	9—10	Embryonic shield
6	11—12	Extraembryonic mesoderm
7	13	Primitive streak
8	14.5	Neural plate

Based on Bryden[31] and Robinson.[33]

X. SHEEP

Carnegie Stages of Sheep Embryos *[Ovis aries]*

The following description of sheep embryos is taken from the works of Bryden,[31] Bryden et al.,[32] and Robinson.[33] In Bryden's studies early embryos up to day 20 were obtained from Columbia-Lincoln cross ewes which had been mated with Suffolk rams. Copulation times were known from direct observation, and the age of each embryo was taken from the time of mating (pc age). The older embryos were from the Cornell University Embryological Collection of Domestic Animals and were of unknown age, breed, and history. The estimation of the age of these embryos is detailed by Bryden et al.[32]

Bryden[31] presents a daily description of organ development in the sheep from days 14 to 34. When this data is examined in terms of Carnegie staging, it becomes obvious that many of these descriptions are of embryos at Carnegie stage boundaries. Nevertheless, a reasonable approximation of the Carnegie staging of sheep embryos is possible. It must be kept in mind that although Bryden[31] presented a reasonably detailed description of internal development, the description was not as complete for the external development, particularly in the development of the footplates. Hence, the following description of Carnegie stages in sheep embryos is to a certain degree tentative. Carnegie stages 1 to 8 of sheep embryos are presented in Table 11.

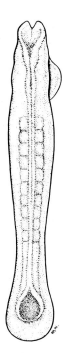

FIGURE 100.   Dorsal view of a Carnegie stage 10 sheep embryo. (From Bryden, M. M., Evans, H. E., and Binns, W., *J. Morph.*, 138, 169, 1972. With permission.)

Sheep. Carnegie Stages 9 and 10

Stage 9 embryos are found on the latter part of day 14 and early day 15 of gestation and have a length of about 2.0 mm. One to three pairs of somites are present, and they have an open neural plate.

Stage 10 embryos are found on the latter part of day 15 and the early part of day 16 of gestation and have a length of 3.0 to 4.0 mm. Four to twelve pairs of somites are present, fusion of the neural folds is occurring in the future neck region, and the otic placode is forming.

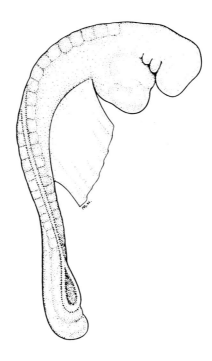

FIGURE 101.   Right aspect of a Carnegie stage 12 sheep embryo. (From Bryden, M. M., Evans, H. E., and Binns, W., *J. Morph.,* 138, 169, 1972. With permission.)

Sheep. Carnegie Stages 11 and 12

Stage 11 begins during the latter part of day 16 and terminates by the end of day 17 of gestation. The embryos range in length from 4.0 to 6.0 mm and have 13 to 20 pairs of somites. This stage is characterized by closure of the cranial neuropore and the appearance of the olfactory placodes. The first and second branchial bars are present. The anlagen of the thyroid gland and liver are present.

Stage 12 embryos are found during day 18 and the early part of day 19 of gestation and range in length from 4.0 to 6.0 mm. There are 21 to 29 pairs of somites. The caudal neuropore closes, the third branchial bar appears, and the yolk stalk is formed. The otic placodes are invaginating. Optic vesicles, lung bud, and Rathke's pouch are seen. The mesonephric ducts fuse with the cloaca.

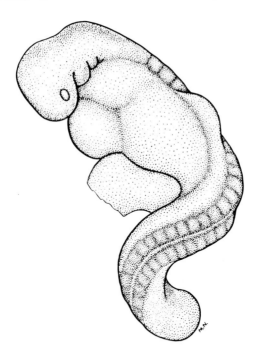

FIGURE 102.   Left aspect of a Carnegie stage 13 sheep embryo. (From Bryden, M. M., Evans, H. E., and Binns, W., *J. Morph.*, 138, 169, 1972. With permission.)

Sheep. Carnegie Stage 13

Stage 13 embryos are found during days 19 and 20 of gestation and range in length from 5.0 to 7.0 mm. These embryos have 30 to 34 pairs of somites and the following external characteristics: closed caudal neuropore, division of the first branchial bar into mandibular and maxillary processes, a distinct cranial limb bud, and closure of the otic vesicle. The caudal limb bud is just beginning to appear. Internal characteristics include: a distinct Rathke's pouch, a definite ventral pancreatic bud, separation of the bilobed lung bud groove from the esophagus, the fourth aortic arch artery, and endocardial cushions.

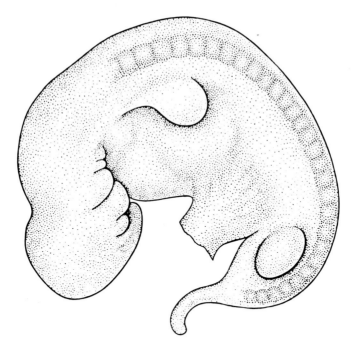

FIGURE 103. Left aspect of a Carnegie stage 15 sheep embryo. (From Bryden, M. M., Evans, H. E., and Binns, W., *J. Morph.*, 138, 169, 1972. With permission.)

Sheep. Carnegie Stages 14 and 15

Stage 14 embryos are found on days 20 and 21 of gestation and range in length from 7.0 to 8.0 mm. There are 35 to 39 pairs of somites, and they have the following external characteristics: distinct projecting cranial and caudal limb buds, distinct cerebral primordia, beginnings of the pontine flexure, and the third and fourth branchial bars are sinking into the cervical sinus. Internal characteristics include: disappearance of the sixth aortic arch artery, complete separation of trachea and esophagus, the ureteric diverticulum, anorectal septum, beginning of the septum primum, interventricular septum, gonad ridges, thinning of the roof of the fourth ventricle, and the optic cup.

Stage 15 embryos are found on day 22 of gestation and range in length from 8.0 to 10.0 mm. The footplate of the cranial limb bud is being delineated, the olfactory placodes are invaginating, and the lens vesicles are closed off. Internal characteristics include: retinal pigmentation, primitive sex cords in the gonads, and secondary budding of the bronchi.

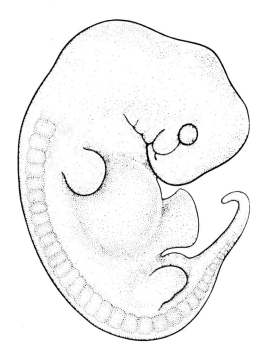

FIGURE 104.   Right aspect of a Carnegie stage 17 sheep embryo. (From Bryden, M. M., Evans, H. E., and Binns, W., *J. Morph.*, 138, 169, 1972. With permission.)

Sheep. Carnegie Stages 16 and 17

Stage 16 embryos are found on day 23 of gestation and range in length from 9.0 to 10.0 mm. They have a distinct paddle-shaped footplate on the cranial limb bud, but not on the caudal limb bud. Internal characters include: tertiary bronchial buds, cochlear duct, completion of septum primum, and almost complete occlusion of the cavity of the optic cup.

Stage 17 sheep embryos are found on days 24 and 25 of gestation and range in length from 11.0 to 13.0 mm. Characteristic external features include: indistinct thoracic somites, toe rays begin to appear on the forelimb footplate, distinct paddle-shaped hindlimb footplate, auricular hillocks, nasomaxillary groove, and fusion of the mandibular processes. Internal characteristics include: lobar and segmental bronchi, septum secundum, disappearance of the dorsal aorta between the third and fourth branchial arch arteries, a small blunt pineal evagination, vomeronasal groove, occlusion of the lens cavity and formation of primary lens fibers, spleen, and chondrification of vertebrae.

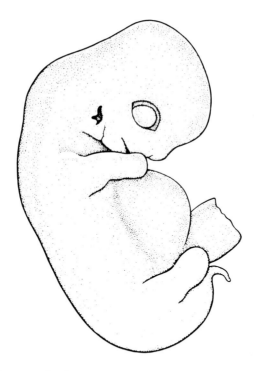

FIGURE 105. Left aspect of a Carnegie stage 18 sheep embryo. (From Marrable, A. W., *The Embryonic Pig. A Chronological Account*, Pitman Medical, London, 1971. With permission.)

Sheep. Carnegie Stage 18

Carnegie stage 18 embryos are found on days 25 to 26 of gestation and range in length from 13.0 to 15.0 mm. The lumbosacral somites are indistinct, and the back is beginning to straighten. There are toe rays on the forelimb footplate. Other external features include: primordium of the genital tubercle, beginning eyelids, beginning closure of the nasolacrimal groove, and merging of the auricular hillocks. Internal features include: chondrification of Meckel's cartilage, humerus, radius, and ulna. There is completion of occlusion of the lens cavity and completion of the interventricular septum. The epithelial bud of the submandibular gland is present.

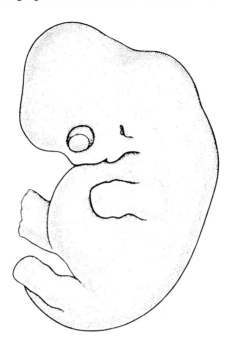

FIGURE 106.   Left aspect of a Carnegie stage 19 sheep embryo. (From Bryden, M. M., Evans, H. E., and Binns, W., *J. Morph.,* 138, 169, 1972. With permission.)

Sheep. Carnegie Stage 19

Stage 19 embryos are found on days 27 and 28 of gestation and range in length from 15.0 to 17.0 mm. Body segmentation is no longer visible externally. There are definite eyelids and a distinct projecting pinna. Pronounced toe rays for the 3rd and 4th digits in the forelimb footplate and a distinct elbow are present. Fusion of the maxillary and nasal processes begins to close the nasolacrimal groove, and distinct nostrils are forming. Internal features include: collecting tubules in the metanephros, rupture of the bucconasal membrane, and chondrification of the maxilla.

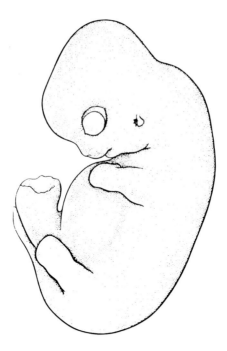

FIGURE 107. Left aspect of a Carnegie stage 20 sheep embryo. (From Marrable, A. W., *The Embryonic Pig. A Chronological Account*, Pitman Medical, London, 1971. With permission.)

Sheep. Carnegie Stage 20 and 21

Stage 20 embryos are found on days 29 and 30 of gestation and range in length from 17.0 to 19.0 mm. More of the eyeballs are covered by the eyelids; the rostrally projecting pinna covers about one half of the external auditory meatus. The spatula-shaped caudal footplate shows toe rays, but rays 3 and 4 are more prominent than rays 2 and 5. There is complete closure of the nasolacrimal groove. The tongue protrudes from the mouth. Internal features include: metanephric vesicles are beginning to form; coiling of the cochlear duct begins; and the ethmoid and nasal bones are chondrified.

Extrapolation of age and size would suggest that stage 21 sheep embryos would be found on days 31 and 32 of gestation and would range in size between 20.0 and 22.0 mm.

FIGURE 108.   Right aspect of a Carnegie stage 22 sheep embryo. (From Marrable, A. W., *The Embryonic Pig. A Chronological Account,* Pitman Medical, London, 1971. With permission.)

Sheep. Carnegie Stage 22

Bryden[31] and Bryden et al.,[32] present very little information applicable to Carnegie staging sheep embryos between days 31 and 34 of gestation. They regard day 34 of gestation as marking the end of the embryonic period; but this is not so, since they categorically state that ossification has not begun. This embryo matches up with Carnegie stage 22, although the relevant data is very sparse. Digits 3 and 4 are separating on the forelimb footplate; the submandibular gland is branching, and its duct is canalizing. The anterior chamber of the eye is present, and the dermocranium consists only of condensed mesenchyme. The paramesonephric duct has reached halfway to the urogenital sinus. The embryo is 24.0 mm long. Since the above stage 22 is given an estimated age of about 33 days, it is likely that a stage 23 sheep embryo would be about 36 days old and about 26.0 mm long.

Table 12
CARNEGIE STAGES 1 TO 8 OF
PIG EMBRYOS

Carnegie stage	PC age (days)	Features
1	1	Fertilization
2	2—3.5	2 — 16 cells
3	4—7	Free blastocyst
4	7—8	Bilaminar embryo
5	8	—
6	9	Primitive streak
7	10—12	Notochordal process
8	13	Neural folds

Based on Marrable.[35]

XI. PIG

Carnegie Stages of Pig Embryos *[Sus scrofa]*

The following description of pig embryos is taken from the work of Paten[34] and Marrable.[35] Patten's account of pig development is based on the Carnegie Institute collection and the data of Minot.[36] He records the postcopulatory age of the various embryos. Marrable[35] cautions about the difficulty of accurately aging embryos and notes that errors of aging may range from 20 to 60% at 5 days, and from 10 to 30% at 10 days. He also gives an equation for estimating age from length and presents a key for the preliminary estimation of pig embryos from their external appearance. Carnegie stages 1 to 8 of pig embryos are presented in Table 12.

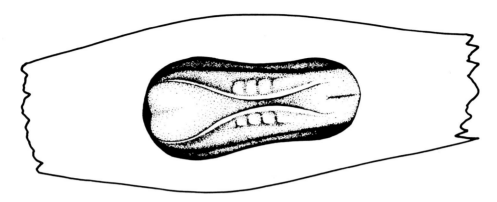

FIGURE 109.   Dorsal view of a Carnegie stage 9 pig embryo. (From Marrable, A. W., *The Embryonic Pig. A Chronological Account,* Pitman Medical, London, 1971. With permission.)

Pig. Carnegie Stage 9

Stage 9 embryos are found on the 14th day of gestation and are about 2.5 mm long. They have one to three pairs of somites and an open neural plate.

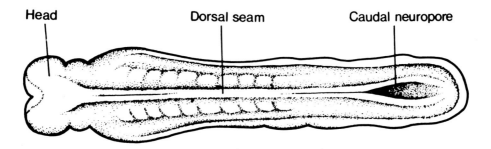

Head   Dorsal seam   Caudal neuropore

FIGURE 110.   Dorsal view of a Carnegie stage 10 pig embryo. (From Marrable, A. W., *The Embryonic Pig. A Chronological Account,* Pitman Medical, London, 1971. With permission.)

Pig. Carnegie Stage 10

Stage 10 embryos are found on the 15th day of gestation and are about 3.0 mm long. There are 4 to 12 pairs of somites present. The neural tube is closing, and the otic placodes are present.

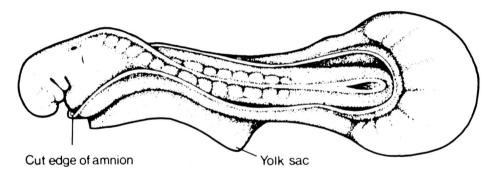

Cut edge of amnion   Yolk sac

FIGURE 111.   Dorsal view of a Carnegie stage 11 pig embryo. (From Marrable, A. W., *The Embryonic Pig. A Chronological Account,* Pitman Medical, London, 1971. With permission.)

Pig. Carnegie Stage 11

Stage 11 embryos are found on the 16th day of gestation and are about 4.5 mm long. There are 13 to 20 pairs of somites present, and towards the end of this stage the cranial neuropore closes. The otic placodes are invaginating, and the mandibular and hyoid bars are present. Internally they are characterized by the hepatic diverticulum, the mesonephros, mesonephric duct, optic vesicles, and the first and second aortic arch arteries.

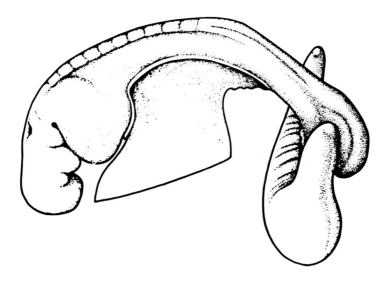

FIGURE 112. Left lateral view of a Carnegie stage 12 pig embryo. Note the large crescentic allantois. (From Marrable, A. W., *The Embryonic Pig. A Chronological Account,* Pitman Medical, London, 1971. With permission.)

Pig. Carnegie Stage 12

Stage 12 embryos are found on the 17th day of gestation and are about 5.0 mm long. They have 21 to 29 pairs of somites and are beginning to undergo torsion. The caudal neuropore closes during this stage. They have a large crescentic allantois; a yolk sac stalk and three branchial bars are now present. Internally they are characterized by the presence of Rathke's pouch, the laryngotracheal groove, the thyroid diverticulum, and the third aortic arch artery. The buccopharyngeal membrane has ruptured.

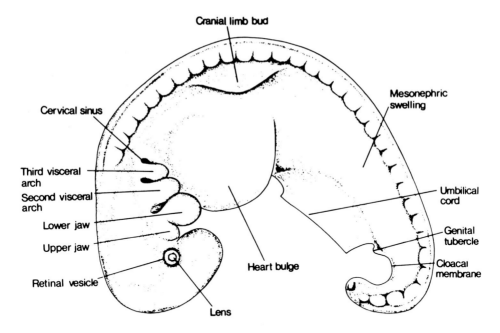

FIGURE 113.   Left lateral view of a Carnegie stage 13 pig embryo. (From Marrable, A. W., *The Embryonic Pig. A Chronological Account*, Pitman Medical, London, 1971. With permission.)

Pig. Carnegie Stage 13

Stage 13 embryos are found on the 18th day of gestation and, because of their greatly increased ventral curvature, they have a crown rump length of about 4.0 mm. There are 30 to 34 pairs of somites present, and there is a distinct cranial limb bud. A definite cervical sinus is present. The first branchial bar is now clearly divided into mandibular and maxillary processes. The lens placode is present. Internally they show a dorsal pancreatic diverticulum, the beginnings of the gallbladder, the endolymphatic duct, and the fourth aortic arch artery.

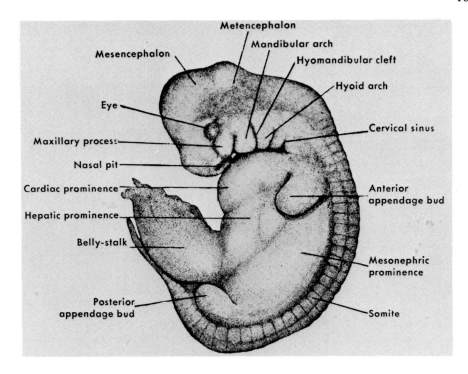

FIGURE 114.   Left lateral view of a Carnegie stage 14 pig embryo. (From Patten, B. M., *Embryology of the Pig*, McGraw-Hill, New York, 1948. With permission.)

Pig. Carnegie Stage 14

Stage 14 embryos are found on the 19th day of gestation and range from 6.0 to 8.0 mm in crown rump length. They have 35 to 39 pairs of somites. The cranial and caudal limb buds are prominent, and there is a distinct nasal pit. The cerebral vesicles are appearing. Internally they are characterized by the optic cup, invagination of the lens placode, a metanephric diverticulum, the endolymphatic duct, the beginning of the tongue, and the sixth aortic arch artery.

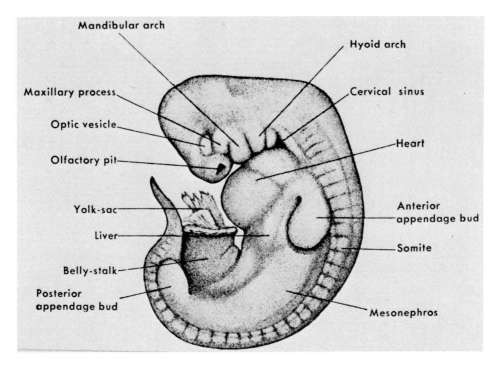

FIGURE 115.   Left lateral view of a Carnegie stage 15 pig embryo. (From Patten, B. M., *Embryology of the Pig,* McGraw-Hill, New York, 1948. With permission.)

Pig. Carnegie Stage 15

Stage 15 embryos are found on days 20 to 21 of gestation and are about 8.0 to 10.0 mm long. The cerebral hemispheres are now well developed. The cranial limb bud is beginning to subdivide into the distal foot segment and a proximal arm segment. There is a pontine flexure, a nasolacrimal groove, and deep olfactory pits. Internally they are characterized by secondary bronchi, the spleen, the aortico-pulmonary septum and the pelvis of the ureter. The midgut is herniated into the root of the umbilical cord.

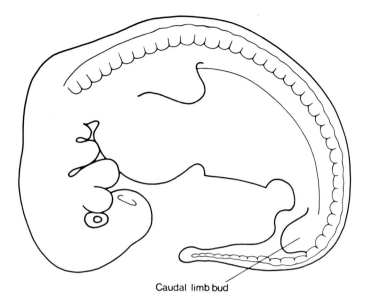

Caudal limb bud

FIGURE 116.   Left lateral view of a Carnegie stage 16 pig embryo. (From Marrable, A. W., *The Embryonic Pig. A Chronological Account,* Pitman Medical, London, 1971. With permission.)

Pig. Carnegie Stage 16

Stage 16 embryos are found on days 21 to 22 of gestation and are about 10.0 to 12.0 mm long. Comparison with other mammalian embryos indicates that at this stage of development pig embryos should have a distinct rounded forelimb footplate, a subdivided caudal limb bud, and auricular hillocks on the second branchial bar. These features are not apparent in the figure from Marrable.[35] The mandibular processes are beginning to fuse, the nasolacrimal groove is distinct, and the olfactory pits are now very deep. The cervical sinus is closing, and the umbilical hernia is more obvious.

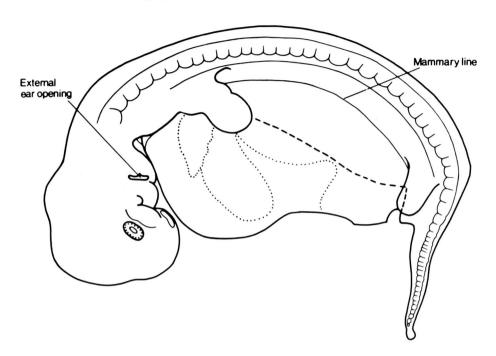

FIGURE 117.   Left lateral view of a Carnegie stage 17 pig embryo. (From Marrable, A. W., *The Embryonic Pig. A Chronological Account,* Pitman Medical, London, 1971. With permission.)

Pig. Carnegie Stage 17

Stage 17 embryos are found on the 23rd day of gestation and are 12.0 to 14.0 mm long. Toe rays are present in the forelimb, and there is a distinct footplate on the hindlimb. There is a distinct vomeronasal groove. Internally there is an olfactory bulb, a neurohypophyseal evagination, beginnings of the semicircular canals, and a pineal evagination. The atrioventricular canal is divided, and the right dorsal aorta between aortic arch arteries three and four has disappeared. Ossification is beginning in the vertebral bodies.

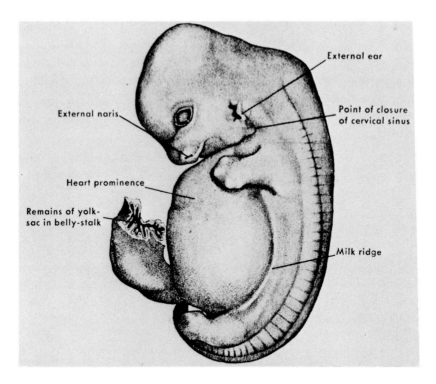

FIGURE 118.   Left lateral view of a Carnegie stage 18 pig embryo. (From Patten, B. M., *Embryology of the Pig,* McGraw-Hill, New York, 1948. With permission.)

Pig. Carnegie Stage 18

Stage 18 embryos are found on the 24th day of gestation and are 14.0 to 16.0 mm long. There are distinct toe rays and the beginning of crenation on the forelimb foot-plate. The auricular hillocks are merging to form the pinna. The milk ridge is present, and the nasolacrimal groove is closing.

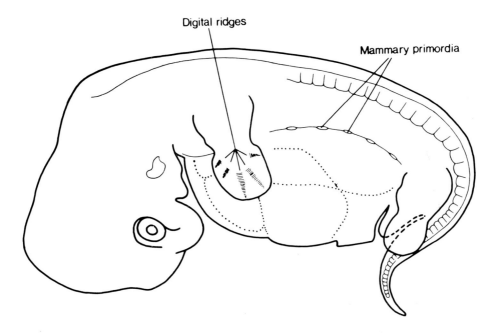

FIGURE 119.   Left lateral view of a Carnegie stage 19 pig embryo.(From Marrable, A. W., *The Embryonic Pig. A Chronological Account*, Pitman Medical, London, 1971. With permission.)

Pig. Carnegie Stage 19

Stage 19 embryos are found on the 25th to 26th day of gestation and are 16.0 to 19.0 mm long. There are toe rays on the hindlimb footplate, the pinna is forming, and there are definite eyelids. The bucconasal membrane has ruptured, and the urorectal septum has divided the cloaca to form the rectum and urogenital sinus.

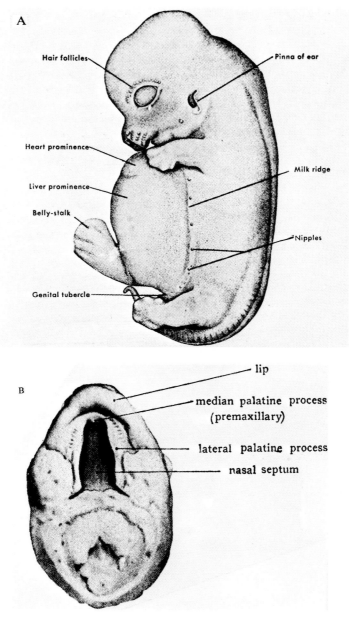

FIGURE 120.   (A) Left lateral view of a Carnegie stage 20 pig embryo. (B) A view of the roof of the mouth of a Carnegie stage 20 pig embryo showing the widely open palate. (From Patten, B. M., *Embryology of the Pig*, McGraw-Hill, New York, 1948. With permission.)

Pig. Carnegie Stages 20, 21, and 22

Stage 20 embryos are found on the 27th and 28th days of gestation and are 19.0 to 22.0 mm long. Hair follicles of the eyebrow are present, as well as developing vibrissae on the snout and cheek. The pinna covers half of the external auditory meatus. The fore- and hindlimbs are almost parallel, with slight flexures at the elbows and knees. The margin of the hindlimb footplate is becoming crenated. The nasolacrimal duct is completely closed, but the secondary palate is wide open.

Stage 21 embryos found on approximately the 29th day of gestation would be about 22.0 to 24.0 mm long. Stage 22 embryos found on about days 30 to 31 of gestation would be about 25.0 to 28.0 mm long. No illustrations of these two stages are available.

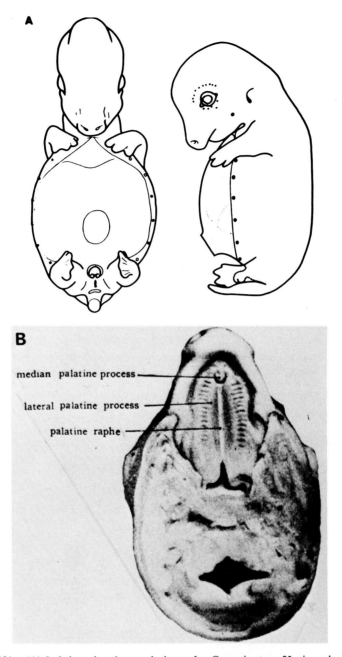

FIGURE 121. (A) Left lateral and ventral views of a Carnegie stage 23 pig embryo. (From Marrable, A. W., *The Embryonic Pig. A Chronological Account,* Pitman Medical, London, 1971. With permission.) (B) A view of the roof of the mouth of a Carnegie stage 23 pig embryo showing closure of the palate. (From Patten, B. M., *Embryology of the Pig,* McGraw-Hill, New York, 1948. With permission.)

Pig. Carnegie Stage 23

Stage 23 embryos are found on days 32 to 33 of gestation and are 28.0 to 31.0 mm long. The digits are beginning to become separated, and the secondary palate is closed.

XII. TREE SHREW

Carnegie Stages of Tree Shrew Embryos *[Tupaia Belangeri]*

Kuhn and Schwaier[37] described 10 stages of the development of the tree shrew *[Tupaia belangeri]*. Insemination was verified by examining vaginal smears, and the postconceptional (pc) age was recorded. The primitive streak and first somites are found in embryos aged 9 days pc. Our staging of the later embryos differs slightly from that put forward by Kuhn and Schwaier.

Tree Shrew. Carnegie Stage 9 (Kuhn and Schwaier Stage IV)

A stage 9 embryo is 1.3 mm long and is aged 9 days pc. It has four pairs of somites, a neural groove, Hensen's node, and a primitive streak.

Tree Shrew. Carnegie Stage 10 (Kuhn and Schwaier Stage V)

A stage 10 embryo is 1.5 mm long and is aged 12 days pc. It has 11 pairs of somites. There is a well-marked "lordosis". Closure of the neural tube has begun.

Tree Shrew. Carnegie Stage 13 (Kuhn and Schwaier Stage VII)

A stage 13 embryo is 4.2 mm long and is aged 14 days pc. It has 31 pairs of somites. The "lordosis" has been reversed, and the embryo is now C-shaped. Branchial bars 1 to 3 are present, and the first is divided into mandibular and maxillary processes. A deep cervical sinus is present. The caudal neuropore is closed. The optic vesicle appears to be separated from the surface ectoderm. The cranial limb bud is very distinct, and the caudal limb bud is just forming. The heart bulge is large.

Tree Shrew. Carnegie Stage 16 (Kuhn and Schwaier Stage VIII)

A stage 16 embryo is 5.2 mm long and is aged 18 days pc. Nasal pits are visible, and there is some pigment in the retina. The cervical flexure is about 90°, and the trunk is less strongly curved. A digital plate is appearing on the cranial limb bud, but the caudal limb bud is still undivided.

Tree Shrew. Carnegie Stage 20 (Kuhn and Schwaier Stage IX)

A stage 20 embryo is 13.6 mm long and aged 24 days pc. A snout with external nares and the primordium of a rhinarium is prominent. Eyelids and auricles are developing. The neck is beginning to appear. The beginning digits are seen on both fore- and hindlimbs. There is a definite tail. The vibrissae primordia are visible.

Tree Shrew. Carnegie Stage 23 (Kuhn and Schwaier Stage X)

A stage 23 embryo is 24.0 mm long and is aged 29 days pc. The eyelids are completely closed suggesting that this is either a late stage 23 or an early fetus. The digits are well formed on both limbs. There are hair follicles over the head and body. The head is well off the chest, and the tail is long and pointed.

XIII. CHICK

Carnegie Stages of Chick Embryos *[Gallus Gallus]*

Streeter[6-10] showed how, in human embryos, the tissues and organs develop in an integrated manner, and this series of events can be used to stage embryos without consideration of their age or length. The various mammalian embryos described in this atlas follow an almost identical sequence of development and can be placed in the same series of Carnegie stages, irrespective of age or length. The anatomy of birds is a marked variation from the basic tetrapod body plan, but in most respects their development is very similar to that of mammals. Hence, chick embryos can be related to the Carnegie stages of mammalian embryos.

Hamburger and Hamilton[38] described 35 stages of chick development covering the entire incubation period and based on external features. Details of the internal features of chick development were obtained from the works of Lillie,[39] Spector,[40] Romanoff,[41] Hamilton,[42] and Freeman and Vince.[43] The beginning of incubation is considered as day zero.

It must be kept in mind that only some of the description of internal development of the chick is referenced to Hamburger and Hamilton stages. The majority of the work is referenced to days of incubation, which we have attempted to relate to Hamburger and Hamilton stages of development. It must also be kept in mind that it is a relatively simple manner to stage chick embryos during the early stages of development, since much work has been done both on chick and mammalian embryos during these stages. It is, however, more problematical for the later stages, as fewer studies have been made on chick and mammalian embryos; hence our interpretation of the Carnegie staging of the chick embryo must be examined critically (see the legend on Figure 151).

FIGURE 122.   Dorsal view of a Carnegie stage 9 chick embryo. (From Hamburger, V. and Hamilton, H. L., *J. Morph.*, 88, 49, 1951. With permission.)

Chick. Carnegie Stage 9 (Hamburger and Hamilton Stages 7 and 8)

Stage 9 embryos have an incubation age of 23 to 29 hr. They have one to three pairs of somites and an open neural plate.

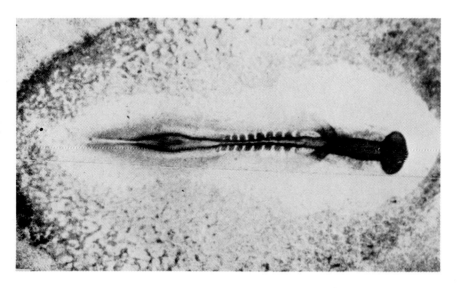

FIGURE 123.   Dorsal view of a Carnegie stage 10 chick embryo. (From Hamburger, V. and Hamilton, H. L., *J. Morph.*, 88, 49, 1951. With permission.)

Chick. Carnegie Stage 10 (Hamburger and Hamilton Stages 8 to 11)

Stage 10 embryos have an incubation age of 29 to 40 hr. During this stage 4 to 12 pairs of somites are present, and closure of the neural tube begins. The otic placode, first aortic arch artery, and bulboventricular loop are present.

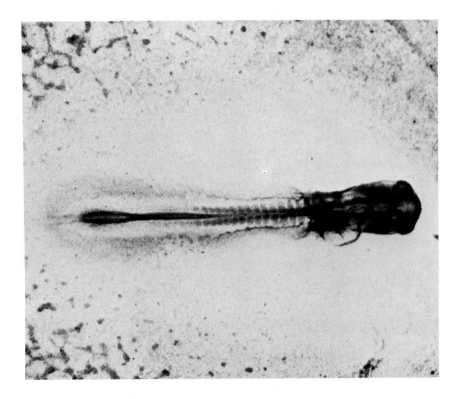

FIGURE 124.   Dorsal view of a Carnegie stage 11 chick embryo. (From Hamburger, V. and Hamilton, H. L., *J. Morph.*, 88, 49, 1951. With permission.)

Chick. Carnegie Stage 11 (Hamburger and Hamilton Stages 11 to 13)

Stage 11 embryos have an incubation age of 40 to 52 hr. There are 13 to 20 pairs of somites present, as well as the mandibular and hyoid bars. Otic pits and optic vesicles are seen. Internal characteristics include the rhombencephalic neural crest, the liver and thyroid outgrowths, and the second aortic arch artery. Closure of the neuropores occurs a little earlier in the chick, since the caudal neuropore closes at the end of stage 10 and the cranial neuropore in stage 12.

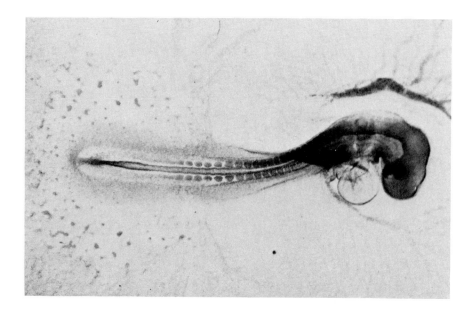

FIGURE 125.   Dorsal view of a Carnegie stage 12 chick embryo. (From Hamburger, V. and Hamilton, H. L., *J. Morph.*, 88, 49, 1951. With permission.)

Chick. Carnegie Stage 12 (Hamburger and Hamilton Stages 13 to 17)

Stage 12 embryos have an incubation age of 50 to 56 hr. They have 21 to 29 pairs of somites and distinct cranial limb buds. A distinct yolk stalk is present. Other distinguishing features include three branchial bars and the olfactory placode. By the end of this stage there is an open otocyst, lens placode, and invaginating optic vesicle. Because of its precocious development, the eye of the chick embryo is not a suitable criterion for Carnegie staging. Internally, embryos of stage 12 have the third aortic arch artery, an interatrial septum, and mesonephric tubules and duct. The bucopharyngeal membrane has ruptured. Rathke's pouch and the lung, liver, and thyroid buds are present.

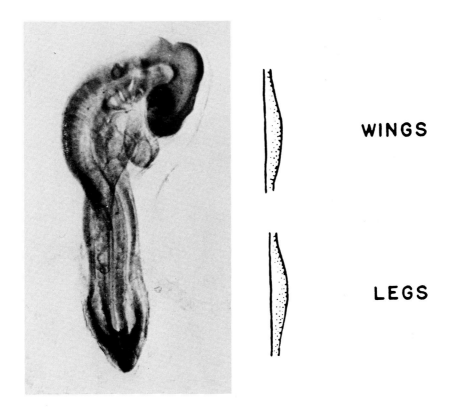

WINGS

LEGS

FIGURE 126.   Right lateral view of a Carnegie stage 13 chick embryo. (From Hamburger, V. and Hamilton, H. L., *J. Morph.*, 88, 49, 1951. With permission.)

Chick. Carnegie Stage 13 (Hamburger and Hamilton Stages 17 and 18)

Stage 13 embryos have an incubation age of 52 to 68 hr. Cranial and caudal limb buds and a distinct tail bud are present. The otocyst is closed. Internally, they have the fourth aortic arch artery, endocardial cushions, the beginnings of the gallbladder, and the dorsal pancreatic outgrowth. The pineal outgrowth appears at this stage, whereas it does not appear until stage 15 in mammals. They have 30 to 34 somites.

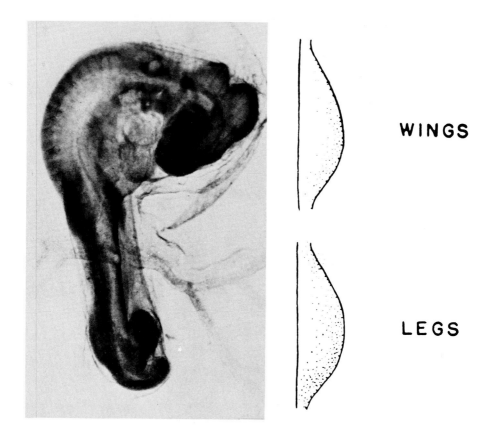

WINGS

LEGS

FIGURE 127.   Right lateral view of a Carnegie stage 14 chick embryo. (From Hamburger and Hamilton, 1951).[38]

Chick. Carnegie Stage 14 (Hamburger and Hamilton Stages 18 and 19)

Stage 14 embryos have an incubation age of 68 to 72 hr. They have 35 to 39 pairs of somites, a distinct maxillary process, a deep olfactory invagination, and the allantois is growing out. The cerebral vesicles and pontine flexure are appearing. The lens vesicle closes at this stage, rather than at stage 15 as in mammals. Internally, they have the sixth aortic arch artery, the interventricular septum and the beginning of cranial nerves IV and XII. The mesonephric duct now joins the cloaca, thus being two stages later than in mammals.

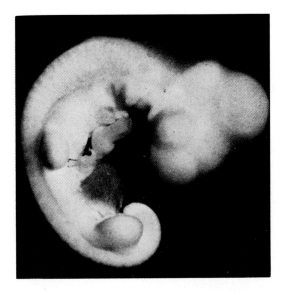

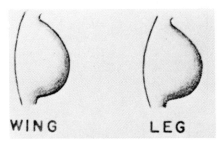

WING   LEG

FIGURE 128.   Right lateral view of a Carnegie stage 15
chick embryo. (From Hamburger, V. and Hamilton, H. L.,
J. Morph., 88, 49, 1951. With permission.)

Chick. Carnegie Stage 15 (Hamburger and Hamilton Stages 20 to 22)

Stage 15 embryos have an incubation age of 3 to 3.5 days. The limb buds project
further, and the olfactory invagination is deeper. The maxillary process is prominent
and is equal to or exceeds the mandibular process in length. Internally, there is retinal
pigment, a cochlear duct, separation of the esophagus and trachea, bronchial buds,
the ventral pancreatic diverticulum, spleen, and the meninx primitiva. At this stage the
ureteric diverticulum, tuberculum impar, adrenal medulla and cortex, eye muscle, and
corneal primordia are also present.

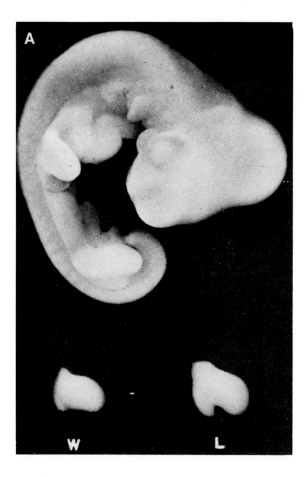

FIGURE 129. (A) Right lateral view of a Carnegie stage 16 chick embryo. W = wing bud; L = leg bud. (B) Right aspect of the branchial bar region. II, III, IV = branchial bars 2, 3, and 4; mx = maxillary process; 4 = 4th branchial cleft; a, b = two protuberances on the mandibular bar; c = receding mandibular process; d, e, f = protuberances on the hyoid bar. (From Hamburger, V. and Hamilton, H. L., *J. Morph.*, 88, 49, 1951. With permission.)

B

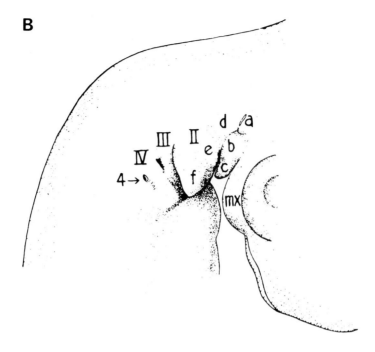

FIGURE 129B.

Chick. Carnegie Stage 16 (Hamburger and Hamilton Stages 23 and 24)

Stage 16 embryos have an incubation age of 3.5 to 4 days. Wing and leg buds are distinctly longer than wider. There is a distinct toe plate in leg bud, but the digital plate is not yet demarcated in the wing bud. Here is a distinct difference between mammals and birds, since in mammals the cranial limb bud is in advance of the caudal limb bud and delineation of the cranial and caudal digital plates occurs slightly earlier. Despite the fact that birds do not have a fleshy pinna, six auricular hillocks appear on the mandibular and hyoid bars at the same time as they do in mammals. Branchial bars three and four are disappearing into the indistinct cervical sinus. The umbilical hernia is present, and the medial and lateral nasal processes are beginning to fuse with the maxillary process. Internally they are characterized by beginning fusion of the endocardial cushions, the appearance of the secondary palatal processes, and the sympathetic chains. The diencephalon is subdividing, and the rhombomeres are becoming indistinct.

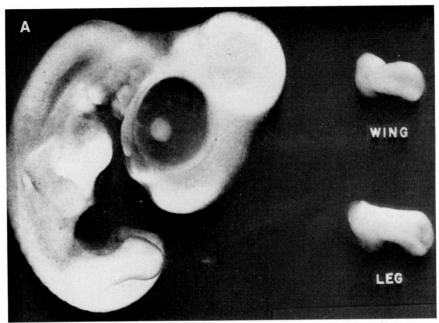

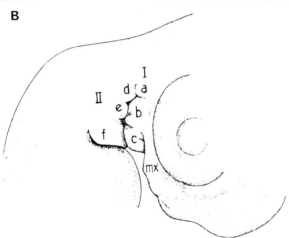

FIGURE 130.   (A) Right lateral aspect of a Carnegie stage 17 chick embryo. (B) Right aspect of the branchial bar region. I and II = mandibular and hyoid bars; mx = maxillary process; a to c = auricular hillocks; f = the "collar". (From Hamburger, V. and Hamilton, H. L., *J. Morph.*, 88, 49, 1951. With permission.)

Chick. Carnegie Stage 17 (Hamburger and Hamilton Stages 25 and 26)

Stage 17 embryos have an incubation age of 4.5 to 5 days. The third and fourth branchial bars have disappeared, the maxillary processes have fused with the medial and lateral nasal processes, and the thoracic somites are still visible. The leg bud is noticeably longer than the wing bud. The elbow and knee joints are seen, and toe rays are beginning to appear in the footplate. The most ventral auricular hillock on the hyoid bar is conspicuous and projects over the surface, and Hamburger and Hamilton[38] refer to it as the "collar". Internal features include: primary sex cords, aortico-pulmonary septum, fusion of the endocardial cushions, chondrification of the humerus, increased size of the secondary palatal folds, and the appearance of the semicircular canals, the posterior commissure, and the primordia of apparently rudimentary vomeronasal organs.

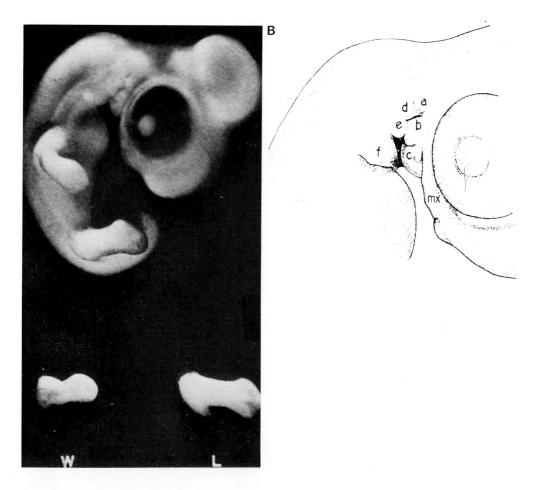

FIGURE 131.   (A) Right lateral aspect of a Carnegie stage 17 chick embryo; W = wing; L = leg. (B) Right aspect of the branchial bar region; a to e = auricular hillocks; f = the "collar"; mx = maxillary process. (From Hamburger, V. and Hamilton, H. L., *J. Morph.*, 88, 49, 1951. With permission.)

Chick. Carnegie Stage 18 (Hamburger and Hamilton Stages 27 and 28)

Stage 18 embryos have an incubation age of 5 to 6 days. The "collar" is more distinct, but the remaining auricular hillocks are beginning to fuse. The toe rays are distinctly demarcated. The lumbosacral somites are indistinct. A prominent outgrowth of the beak is visible at the end of this stage. There is a nasolacrimal groove. Internal features include: three semicircular canals, closure of the lens cavity, olfactory nerves are reaching the olfactory bulb, paramesonephric ducts, chondrogenesis of Meckel's cartilage, further chondrification of the limb bones, closure of the pituitary stalk, and almost complete closure of the interventricular septum.

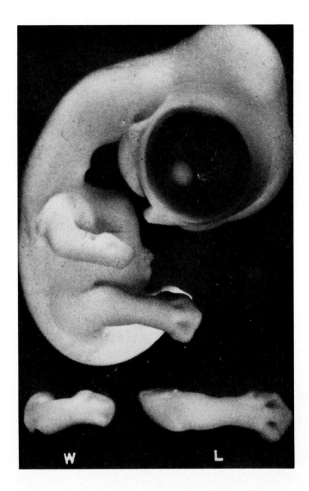

FIGURE 132. Right lateral aspect of a Carnegie stage 19 chick; W = wing; L = leg. (From Hamburger, V. and Hamilton, H. L., *J. Morph.*, 88, 49, 1951. With permission.)

Chick. Carnegie Stage 19 (Hamburger and Hamilton Stages 29 and 30)

Stage 19 embryos have an incubation age of 6 to 6.5 days. The leg is still markedly longer than the wing. Both wing and leg have distinct digital rays. The beak is more prominent, and the egg tooth begins to appear at the end of this stage. The eyelids are distinct, and the external nares are closed. Feather germs are present on the back at the level of the wings and legs. The "collar" is still very distinct, but the other auricular protuberances have flattened and fused to form the very shallow external auditory meatus. The neck has markedly lengthened between the "collar" and the mandible. Scleral papillae are appearing but no more than two at this stage. Some reflexes can be elicited. Internal features include: a solid pituitary stalk, chondrification of the maxilla, the thoracic duct, and the sex of the gonads is apparent at the end of this stage.

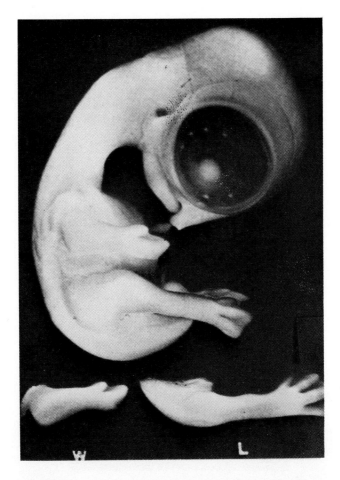

FIGURE 133.   Right lateral aspect of a Carnegie stage 20 chick embryo; W = wing; L = leg. (From Hamburger, V. and Hamilton, H. L., *J. Morph.*, 88, 49, 1951. With permission.)

Chick. Carnegie Stage 20 (Hamburger and Hamilton Stages 31 and 32)

Stage 20 embryos have an incubation age of 7 to 7.5 days. All digits are delineated and considerably lengthened by the end of this stage. The anterior tip of the mandible has reached the beak. The "collar" is disappearing. More feather germs have appeared. Up to eight scleral papillae are arranged in dorsal and ventral groups, but the circle is not yet complete. Internal features include: formation of the nasolacrimal duct, distinct conjunctival sac, beginnings of the ciliary body, junction of the oviduct with the cloaca, chondrification of the otic capsule and nasal septum, and ossification of the clavicle.

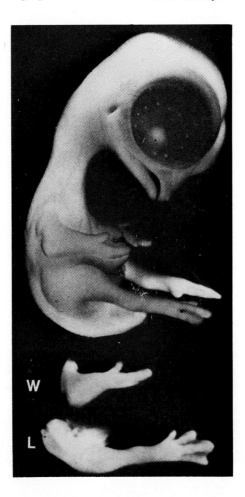

FIGURE 134.   Right lateral aspect of a Carnegie stage 21 chick embryo. W = wing; L = leg. (From Hamburger, V. and Hamilton, H. L., *J. Morph.*, 88, 49, 1951. With permission.)

Chick. Carnegie Stage 21 (Hamburger and Hamilton Stages 33 and 34)

Stage 21 embryos have an incubation age of 7.5 to 8 days. The toes are separating, and the first digit of the wing is distinctly set aside from the others. Webs are visible between the digits. The mandible and neck have lengthened considerably. More feather germs are present. Nine to 14 scleral papillae are present, and the circle is complete by the end of this stage. The nictitating membrane extends halfway between the outer rim of the eye and the scleral papillae. Internal features that place Hamburger and Hamilton stages 33 and 34 in Carnegie stage 21 include completion of the semilunar valves and a distinct circular muscle layer in the gut wall. Ossification is commencing in the long bones.

FIGURE 135.   Right lateral aspect of a Carnegie stage 22 chick embryo; W = wing; L = leg. (From Hamburger, V. and Hamilton, H. L., *J. Morph.*, 88, 49, 1951. With permission.)

Chick. Carnegie Stage 22 (Hamburger and Hamilton Stage 35)

Stage 22 embryos have an incubation age of 8 to 9 days. The digits and toes are further separated, and the webs are now inconspicuous. All feather germs are more conspicuous. The nictitating membrane is closer to the ring of scleral papillae, and eyelids are forming. Internal features include: the appearance of the neostriatum (the avian homologue of the mammalian neopallium) and ossification in all the long bones.

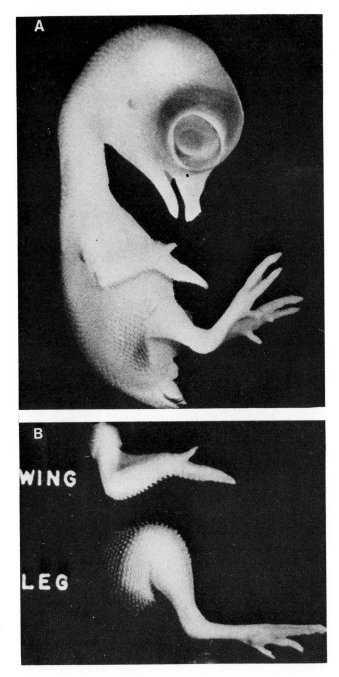

FIGURE 136.   (A) Right lateral aspect of a Carnegie stage 23 chick embryo. (B) Right aspect of wing and leg. (From Hamburger, V. and Hamilton, H. L., *J. Morph.*, 88, 49, 1951. With permission.)

Chick. Carnegie Stage 23 (Hamburger and Hamilton Stage 36)

Stage 23 embryos have an incubation age of 10 days. The secondary palate is closed, and the eyelids are beginning to close. The toes are completely separated, and the claws are appearing. There is a well-defined pia-arachnoid membrane, and many cranial bones are ossifying.

CORRELATION GRAPHS

The following graphs relate crown rump length and developmental age to Carnegie stages 9 to 23 and will aid investigators in staging embryos. However, these correlations are affected by several variable factors and should not be regarded as being absolute. For example, there is a considerable spread of crown rump length within a given Carnegie stage (Figure 2), and, for the most part, the lengths given were made after fixation and are generally less than those of fresh specimens. The differing degree of curvature of embryos from about Carnegie stage 12 onwards influences the crown rump length. As shown by Juurlink and Fedoroff,[5] there may be considerable variation in the stage of development reached by embryos at any particular day of gestation (Figure 3) or incubation.

The postcopulatory or incubation age can only be used as a rough indication of the stage of development because:

- There is uncertainty as to the precise time of fertilization in mammals, particularly in the primates.
- Differences in the time of implantation of the members of a litter is probably a main cause of individual variations in stages of development.
- Two main causes of variation in individual chick development are differing times of passages along the oviduct and how long the eggs have been incubated by a broody hen before collection.

Hence, these graphs can only be a guide to the probable Carnegie stage attained by a particular embryo, and the final precise staging will be made on its external morphological characters.

When mean age vs. Carnegie stage of development is plotted for the placental mammals, the most striking aspect of the resulting curve is that it is linear from Carnegie stages 9 to 23. The slope of the curve when expressed in days reflects the mean time interval occupied by one Carnegie stage. This is interpreted to mean that from Carnegie stages 9 to 23 the rate of development of the embryos is constant, and the varying slopes found among the different species reflects their different rates of development. This constancy in the rate of development is not necessarily true for earlier Carnegie stages. For example, in the mouse the mean length of a Carnegie stage is about 0.5 days, be it stage 7 or 23. In the marmoset the mean length of each stage is about 1.7 days between stages 9 and 23, whereas the length of stage 8 is about 10 days, and the length of stage 7 is about 20 days.[17] Changes in the rate of development between Carnegie stages 7 and 9 of such magnitude have only been encountered in the marmoset; nevertheless, smaller but real changes in the rate of development between Carnegie stages 7 and 9 can also be observed in several other species. This suggests that the embryonic period proper should be considered to begin at Carnegie stage 9, i.e., the beginning of somite formation rather than the appearance of the primitive streak.

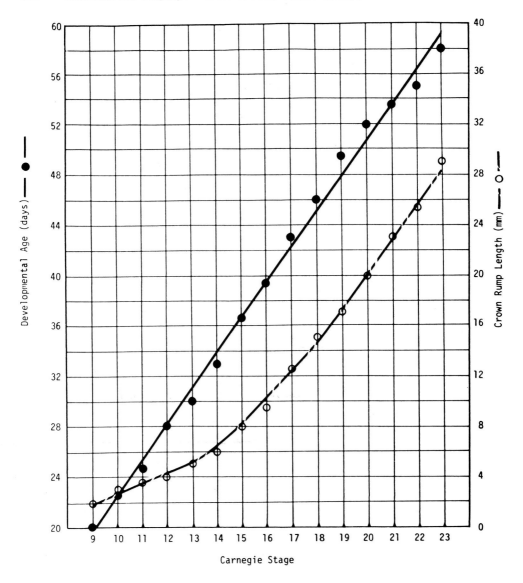

FIGURE 137.   Graphs relating postovulatory age and crown rump length to the Carnegie stages of development of the human embryo.

MAN

This information is taken from Table 1 in the review article by O'Rahilly.[44] The embryonic ages are recorded as estimated postovulatory ages, but the points on the graph are the midpoints of the ranges given by O'Rahilly. The crown rump lengths are those of fixed embryos, and again the midpoints are plotted. A linear regression analysis was performed on mean age vs. Carnegie stage of development, which gave a slope of 2.8321; i.e., each Carnegie stage of human development occupies about 2.8 days.

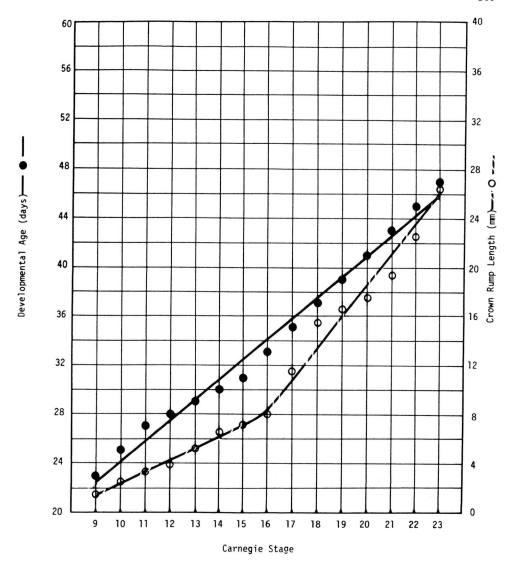

FIGURE 138.   Graphs relating estimated fertilization age and crown rump length to the Carnegie stages of development of the baboon embryo.

BABOON

This information is taken from Hendrickx.[13] The embryonic ages are plotted as the mean estimated fertilization ages as described by Hendrickx. The crown rump lengths are those of fixed embryos, and the midpoints of the ranges given by Hendrickx are plotted. A linear regression analysis was performed on mean age vs. Carnegie stage of development, which gave a slope of 1.6714; i.e., each Carnegie stage of baboon development occupies about 1.7 days.

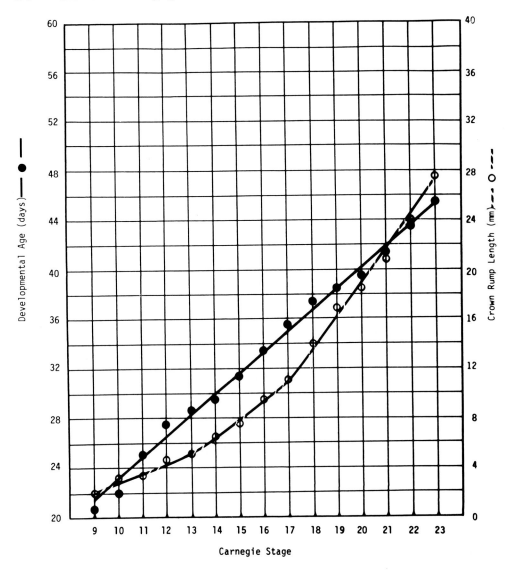

FIGURE 139.   Graphs relating estimated fertilization age and crown rump length to the Carnegie stages of development of the rhesus monkey embryo.

RHESUS MONKEY

This information is taken from Hendrickx and Sawyer.[14] The embryonic ages are recorded as estimated fertilization ages, and the points on the graph are the midpoints of the ranges given by Hendrickx and Sawyer. Similarly, the crown rump lengths are plotted as the midpoints of the ranges given for fixed embryos. A linear regression analysis was performed on mean age vs. Carnegie stage of development, which gave a slope of 1.7304; i.e., each Carnegie stage of rhesus development occupies about 1.7 days.

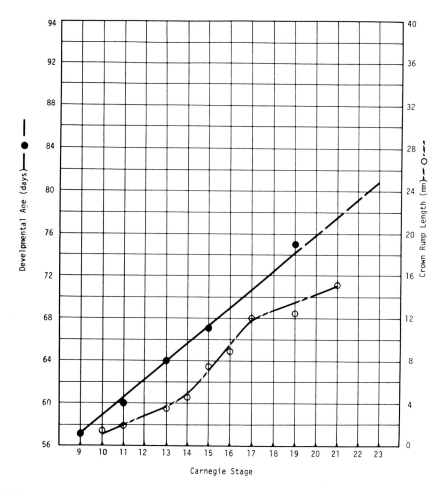

FIGURE 140. Graphs relating estimated fertilization age and crown rump length to the Carnegie stages of development of the common marmoset embryo.

MARMOSET

The method used in breeding these animals was generally by pairing females monogamously with the male before conception and then throughout pregnancy. This information is taken from Phillips,[17] who presented the age of the resulting embryos in terms of minimum, and maximum possible fertilization ages. These estimates generally have rather large ranges. Carnegie stage 9 embryos resulted from a pregnancy with a range of only 4 days between minimum and maximum ages; therefore the mean was used to plot the graph. Embryos at Carnegie stages 13 and 15 were obtained from more than one pregnancy at each stage of development. It was felt that the best estimate of the age of these embryos was for any one stage of development to average the maximal minimum age with the minimal maximum age presented by Phillips. Some embryos at Carnegie stages 11 and 19 of development resulted from a single 24 hr period of mating, and these values were used for plotting these stages. For the remaining stages of development, it was felt that no reasonable estimate of age could be made. Crown rump lengths were from fixed embryos, and the points plotted are the midpoints of the range given by Phillips. Using the five points where there were reasonable estimates of embryonic ages, a linear regression analysis was performed on mean age vs. Carnegie stage of development, which gave a slope of 1.7000; i.e., each Carnegie stage of marmoset development therefore occupies about 1.7 days. The broken line represents the extrapolation of this curve for Carnegie stages 20 to 23.

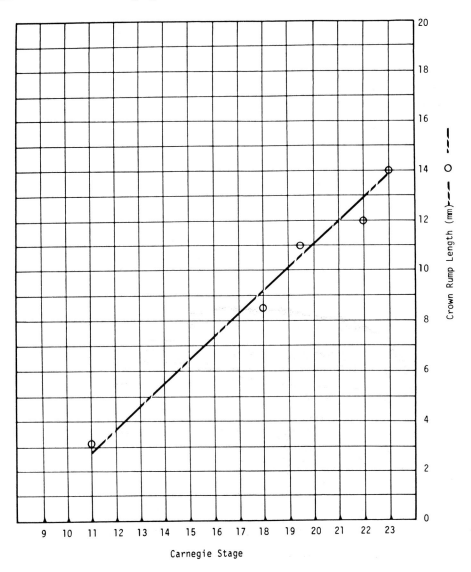

FIGURE 141.   Graph relating crown rump length to the Carnegie stages of development of the lesser galago embryo.

LESSER GALAGO

The crown rump lengths plotted are of fixed embryos, and the data is obtained from the personal observations of H.B. Embryonic age is only known for the stage 23 embryo, which has a postcopulation age of 61 to 62 days.

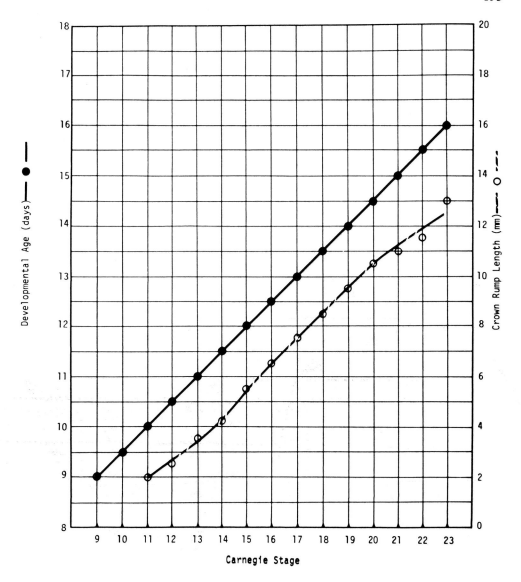

FIGURE 142. Graphs relating postcopulatory age and crown rump length to the Carnegie stages of development of the mouse embryo.

MOUSE

This information is taken from Theiler.[18] The embryonic ages are recorded as postcopulatory ages; i.e., the morning that the vaginal plug is observed is considered day 1 of gestation. The crown rump lengths plotted are the midpoints of the ranges of unfixed embryos. Linear regression analysis was performed on mean age vs. Carnegie stage of development, which gave a slope of 0.5000; i.e., each Carnegie stage of mouse development occupies about 0.5 of a day.

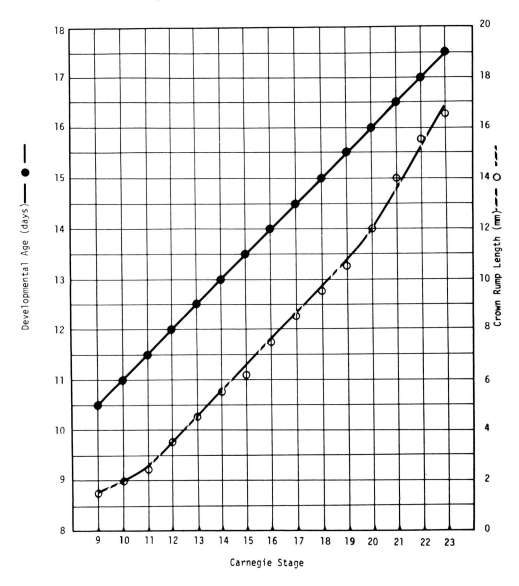

FIGURE 143.   Graphs relating postcopulatory age and crown rump length to the Carnegie stages of development of the rat embryo.

RAT

This information is taken from Witschi[20] and personal observations by B.H.J.J. The crown rump lengths plotted are the midpoints of the ranges of fixed embryos given by Witschi. A linear regression analysis was performed on mean age vs. Carnegie stage of development, which gave a slope of 0.5000; i.e., each Carnegie stage of rat development occupies about 0.5 of a day.

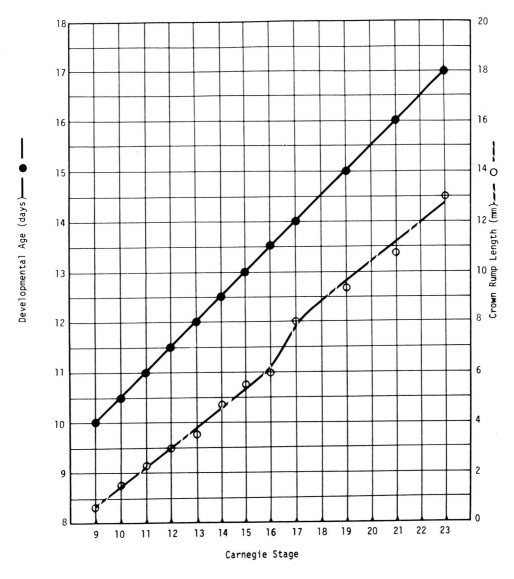

FIGURE 144. Graphs relating postcopulatory age and crown rump length to the Carnegie stages of development of the Chinese hamster embryo.

CHINESE HAMSTER

This information is taken from Donkelaar et al.[21] The crown rump lengths plotted are the midpoints of the ranges of fixed embryos. A linear regression analysis was performed on mean age vs. Carnegie stage of development, which gave a slope of 0.5000; i.e., each Carnegie stage of Chinese hamster development occupies about 0.5 of a day.

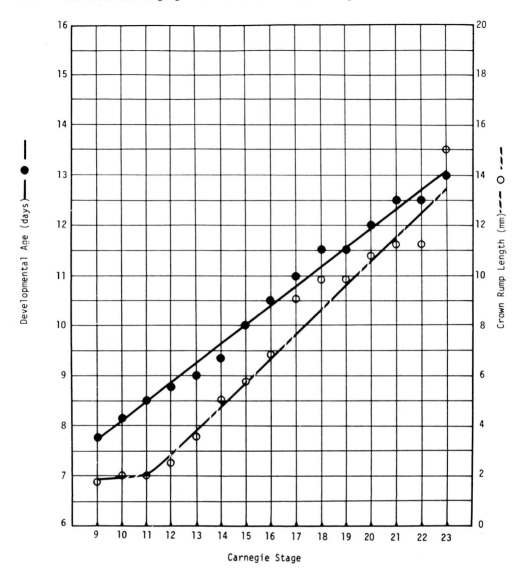

FIGURE 145.   Graphs relating postcopulatory age and crown rump length to the Carnegie stages of development of the Golden hamster embryo.

GOLDEN HAMSTER

This information is taken from Boyer.[23] Both the postcopulatory ages and crown rump lengths plotted are the midpoints of the ranges given by Boyer. A linear regression analysis was performed on mean age vs. Carnegie stage of development, which gave a slope of 0.3884; i.e., each Carnegie stage of golden hamster development occupies about 0.4 of a day.

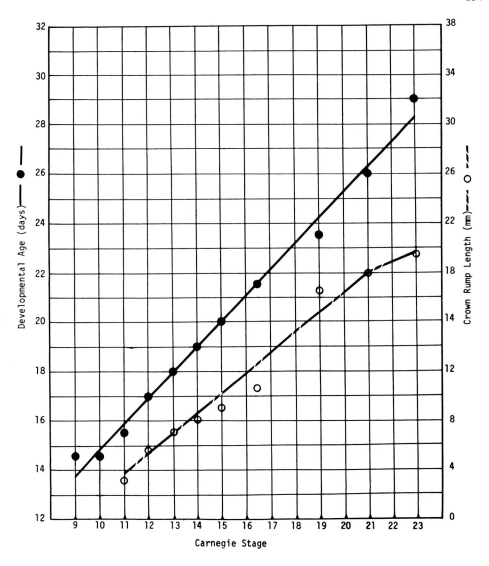

FIGURE 146. Graphs relating postcopulatory age and crown rump length to the Carnegie stages of development of the guinea pig embryo.

GUINEA PIG

This information is taken from Harman and Prickett,[24] Harman and Dobrovolny,[25] and Scott.[26] The crown rump lengths are the midpoints of the ranges of unfixed embryos. A linear regression analysis was performed on mean age vs. Carnegie stage of development, which gave a slope of 1.0359; i.e., each Carnegie stage of guinea pig development occupies about one day.

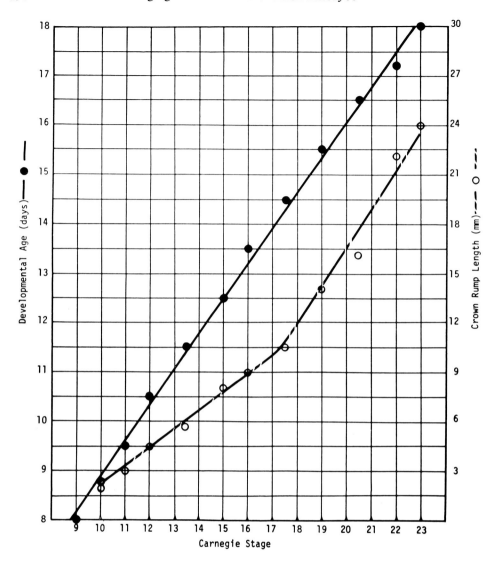

FIGURE 147.   Graphs relating postcopulatory age and crown rump length to the Carnegie stages of development of the rabbit embryo.

RABBIT

This information is taken from Edwards.[29] Crown rump lengths are of fixed embryos and have been obtained directly from the photographs in the paper by Edwards. Linear regression analysis was performed on mean age vs. Carnegie stage of development, which gave a slope of 0.7134; i.e., each Carnegie stage of rabbit development occupies about 0.7 of a day.

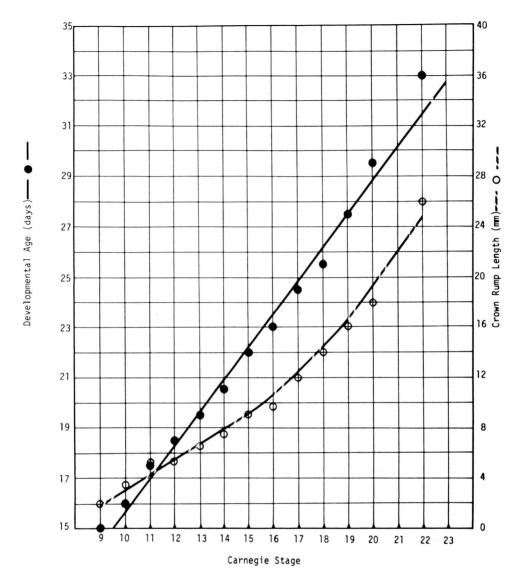

FIGURE 148. Graphs relating postcopulatory age and crown rump length to the Carnegie stages of development of the sheep embryo.

SHEEP

This information is taken from Bryden[31] and Bryden et al.[32] Up to stage 14 of development the embryonic ages are known postcopulatory ages, but from stage 15 onwards the embryonic ages are estimated postcopulatory ages. The crown rump lengths are the midpoints of the ranges of fixed embryos. A linear regression analysis was performed on mean age vs. Carnegie stage of development, which gave a slope of 1.3315; i.e., each Carnegie stage of sheep development occupies about 1.3 days.

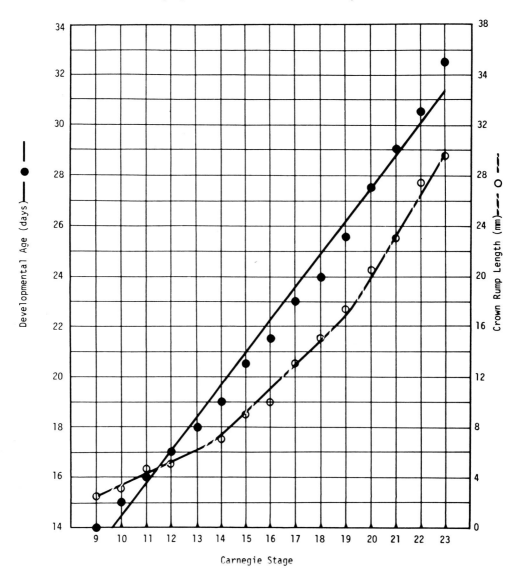

FIGURE 149.   Graphs relating postcopulatory age and crown rump length to the Carnegie stages of development of the pig embryo.

PIG

This information is taken from Marrable.[35] The crown rump lengths are the midpoints of the ranges of fixed embryos A linear regression analysis was performed on mean age vs. Carnegie stage of development, which gave a slope of 1.3714; i.e., each Carnegie stage of pig development occupies about 1.4 days.

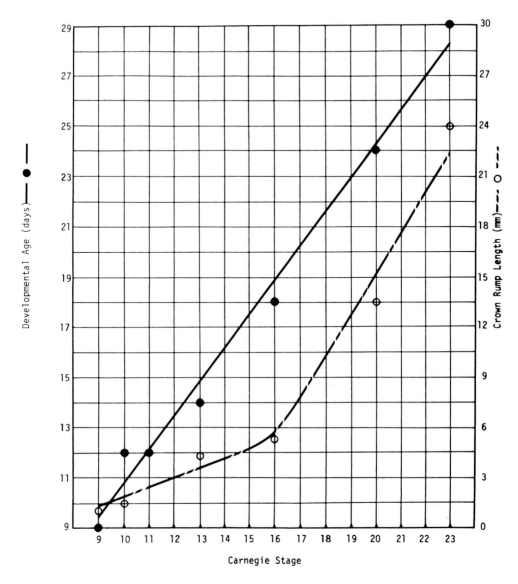

FIGURE 150.   Graphs relating postcopulatory age and crown rump length to the Carnegie stages of development of the tree shrew embryo.

TREE SHREW

This information is taken from Kuhn and Schwaier.[37] The crown rump lengths given are the measurements of only one fixed embryo at each of the stages described by the above authors. A linear regression analysis was performed on mean age vs. Carnegie stage of development, which gave a slope of 1.3586; i.e., each Carnegie stage of tree shrew development lasts about 1.4 days.

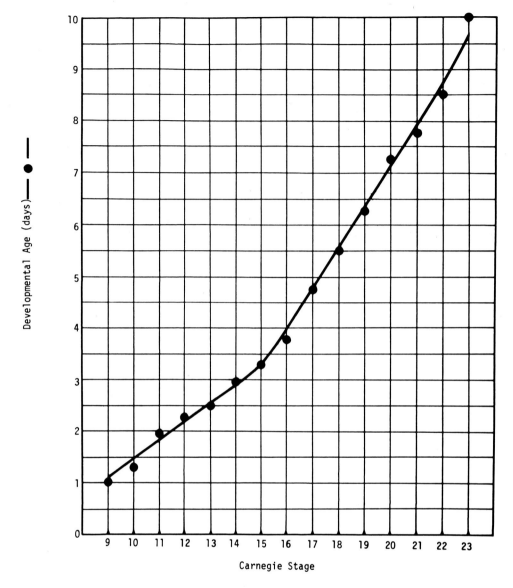

FIGURE 151.   Graph relating incubation age to the Carnegie stages of development of the chick embryo.

CHICK

The basis for this interpretation of Carnegie staging of the chick embryo has been previously outlined. The incubation ages given are for an incubation temperature of about 38°C. When presumptive Carnegie stages are plotted vs. incubation age, Carnegie stage 15 appears to be a pivotal point in the growth rate of the embryo. The curve from Carnegie stage 9 to 15 has a slope of 0.3746, whereas the curve from Carnegie stage 15 to 23 has a slope of 0.8167. This suggests that for the earlier stages of development each Carnegie stage occupies about 0.4 of a day, whereas for the later stages of development each Carnegie stage occupies about 0.8 of a day. How valid is our interpretation of the Carnegie staging of the chick embryo? Since there is a great deal known about the early development of the chick embryo, there is no doubt that our staging is reasonable for this time period. This is the time when the graph has a slope of about 0.4. For the latter part of embryonic development when the graph has a slope of about 0.8, there is much less information available to place chick embryos in Carnegie stages. As a check on the validity of our interpretation of the Carnegie staging of the chick embryo, the following parameters of central nervous system development were compared between the chick and the mammal: first appearance of glioblasts in the spinal cord, first signs of elicited reflexes, the periods of neuronogenesis in the spinal cord, and the formation of Purkinje cells in the cerebellum. In the chick, glioblasts appear in the spinal cord between 6.5 and 8 days of incubation;[45] this would therefore place it at about Carnegie stages 19 to 20. In the mouse, glioblasts appear at about the same stages of development (personal observations, B.H.J.J.). In man, the earliest elicited reflexes occurs at about Carnegie stage 20,[46] whereas in the chick this occurs during Hamilton and Hamburger stages 29 to 30,[43] and thus during Carnegie stage 19 of development. In the mouse, neuronogenesis occurs in the spinal cord from Carnegie stages 11 to 20,[47] whereas in the chick,[48] spinal cord neuronogenesis occurs between the third and eighth day of incubation, which, according to our interpretation of chick development, corresponds to Carnegie stages 11 to 20. Finally, Purkinje cell formation occurs in the chick during the fifth to seventh day of incubation,[49] which would place it from about Carnegie stages 16 to 20, whereas, in the mouse, Purkinje cells are formed during days 12 to 14 postcopulation,[50] thus placing it from about Carnegie stages 15 to 19. It is therefore concluded that our interpretation of the Carnegie staging of the chick embryo is a reasonable one.

REFERENCES

1. Mall, F. P., On stages in the development of human embryos from 2 to 25 mm long, *Anat. Anz.*, 46, 78, 1914.
2. O'Rahilly, R., *Developmental stages in human embryos, including a survey of the Carnegie collections. Part A. Embryos of the first three weeks (Stages 1 to 9)*, Carnegie Institute of Washington, Washington, D.C., 1973.
3. Streeter, G. L., Weight, sitting height, head size, foot length and menstrual age of the human embryo, *Contrib. Embryol. Carneg. Inst.*, 11, 143, 1920.
4. Wilson, J. G., *Environmental and Birth Defects*, Academic Press, New York and London, 1973, 135.
5. Juurlink, B. H. J. and Fedoroff, S., The development of mouse spinal cord in tissue culture. I. Cultures of whole mouse embryos and spinal-cord primordia, *In Vitro*, 15, 86, 1980.
6. Streeter, G. L., Description of age groups XI, 13 to 20 somites, and age group XII, 21 to 29 somites, *Contrib. Embryol. Carneg. Inst.*, 30, 211, 1942.
7. Streeter, G. L., Description of age group XIII, embryos of about 4 or 5 millimeters long and age group XIV, period of indentation of the lens vesicle, *Contrib. Embryol. Carneg. Inst.*, 31, 27, 1945.
8. Streeter, G. L., Description of age groups XV, XVI, XVII and XVIII, being the third issue of a survey of the Carnegie collection, *Contrib. Embryol. Carneg. Inst.*, 32, 153, 1948.
9. Streeter, G. L., A review of histiogenesis of cartilage and bone, *Contrib. Embryol. Carneg. Inst.*, 33, 149, 1949.
10. Streeter, G. L., Description of age groups XIX, XX, XXI, XXII and XXIII, being the fifth issue of a survey of the Carnegie collection, *Contrib. Embryol. Carneg. Inst.*, 34, 165, 1951.
11. Heuser, C. H. and Corner, G. W., Developmental horizons in human embryos. Description of age group X, 4 to 12 somites. *Contrib. Embryol. Carneg. Inst.*, 36, 29, 1957.
12. Gasser, R. F., *Atlas of Human Embryos*, Harper and Row, Maryland, 1975.
13. Hendrickx, A. G., *Embryology of the Baboon*, University of Chicago Press, Chicago, 1971.
14. Hendrickx, A. G. and Sawyer, R. H., Embryology of the rhesus monkey, in *The Rhesus Monkey*, Bourne, G. H., Ed., Academic Press, New York, 1975, 141.
15. Heuser, C. H. and Streeter, G. L., Development of the macaque embryo, *Contrib. Embryol. Carneg. Inst.*, 29, 15, 1941.
16. Gribnau, A. A. M. and Geijsberts, L. G. M., Developmental stages in the Rhesus monkey *(Macaca mulatta)*, *Adv. Anat. Embryol. Cell Biol.*, 68, 1, 1981.
17. Phillips, I. R., The embryology of the common marmoset *(Callithrix jacchus)*, *Adv. Anat. Embryol. Cell Biol.*, 52, 5, 1976.
18. Theiler, K., *The House Mouse*, Springer-Verlag, New York, Heidelberg, Berlin, 1972.
19. Otis, E. M. and Brent, L., Equivalent ages in mouse and human embryos, *Anat. Rec.*, 120, 33, 1954.
20. Witschi, E., Development: Rat, in *Growth*, Altman, P. L. and Dittmer, D. S., Eds., Federation of American Societies for Experimental Biology, Washington, D.C., 1962.
21. Donkelaar, H. J. ten, Geijsberts, L. G. M., and Dederen, P. J. W., Stages in the prenatal development of the Chinese hamster *(Cricetulus griseus)*, *Anat. Embryol.*, 156, 1, 1979.
22. Pickworth, S., Yerganian, G., and Chang, M. C., Fertilization and early development in the Chinese hamster, *Cricetulus griseus, Anat. Rec.*, 162, 197, 1968.
23. Boyer, C. C., Embryology, in *The Golden Hamster*, Hoffman, R. A., Robinson, P. F., and Magalhaes, H., Eds., The Iowa State University Press, Ames, Iowa, 1968, Chap. 5.
24. Harman, M. T. and Prickett, M., The development of the external form of the guinea-pig *(Cavia cobaya)* between the ages of 11 days and 20 days of gestation, *Am. J. Anat.*, 49, 351, 1932.
25. Harman, M. T. and Dobrovolny, M., The development of the external form of the guinea-pig *(Cavia cobaya)* between the ages of 21 days and 35 days of gestation, *J. Morph.*, 54, 493, 1933.
26. Scott, J. P., The embryology of the guinea pig. 1. A table of normal development, *Am. J. Anat.*, 60, 351, 1937.
27. Huber, G. C., On the anlage and morphogenesis of the chorda dorsalis in mammalia, in particular the guinea pig *(Cavia cobaya)*. *Anat. Rec.*, 14, 217, 1918.
28. Hartman, H. A., The fetus in experimental teratology, in *The Biology of the Laboratory Rabbit*, Weisbroth, S. H., Flatt, R. E., and Kraus, A. L., Eds., Academic Press, New York and London, 1974, Chap. 5.
29. Edwards, J. A., The external development of the rabbit and rat embryo, in *Advances in Teratology*, Vol. 3, Woollam, D. H. M., Ed., Academic Press, New York, 1968, Chap. 7.
30. Marshall, A. M., *Vertebrate Embryology*, Smith, Elder and Co., London, 1893, Chap. 5.
31. Bryden, M. M., Prenatal developmental anatomy of the sheep with particular reference to the period of the embryo (11 to 34 days). Thesis for D.Sc. in V.M., Cornell University, New York, 1969.
32. Bryden, M. M., Evans, H. E., and Binns, W., Embryology of the sheep. I. Extraembryonic membranes and the development of body form, *J. Morph.*, 138, 169, 1972.

33. Robinson, T. J., Reproduction in the ewe, *Biological Reviews,* 25, 121, 1951.
34. Patten, B. M., *Embryology of the Pig,* McGraw-Hill, New York, 1948.
35. Marrable, A. W., *The Embryonic Pig. A Chronological Account,* Pitman Medical, London, 1971.
36. Minot, C. S., *A Laboratory Textbook of Embryology,* 2nd ed., Blakiston, Philadelphia, 1911.
37. Kuhn, H-J and Schwaier, A., Implantation, early placentation and the chronology of embryogenesis in *Tupaia belangeri., Z. Anat. Entwickl.-Gesch.,* 142, 315, 1973.
38. Hamburger, V. and Hamilton, H. L., A series of normal stages in the development of the chick embryo, *J. Morph.,* 88, 49, 1951.
39. Hamilton, H. L., *Lillie's Development of the Chick.,* Henry Holt, New York, 1952.
40. Spector, W. S., Prenatal development, various organs and tissues: *Chick. Handbook of Biological Data,* W.B. Saunders and Company, Philadelphia, p. 154.
41. Romanoff, A. L., *The Avian Embryo,* Macmillan Co., New York, 1969.
42. Hamilton, H. L., Development: chick, in *Growth,* Altman, P. L. and Dittmer, D. S., Eds., Federation of American Societies for Experimental Biology, Washington, D.C., 1962.
43. Freeman, B. M. and Vince, M. A., *Development of the avian embryo,* Chapman and Hall, London, 1974, p. 269.
44. O'Rahilly, R., Early human development and the chief source of information on staged human embryos, *Europ. J. Obstet. Gynec. Reprod. Biol.,* 9/4, 273, 1979.
45. Fedoroff, S. and Doering, L. C., Colony culture of neural cells as a method for the study of cell lineages in the developing CNS: The astrocyte lineage, *Curr. Top. Develop. Biol.,* 16, 283, 1980.
46. Hines, M. and Bartelmez, G. W., Development of tissues and organs: Man in *Growth,* Altman, P. L. and Dittmer, D. S., Eds., Federation of American Societies for Experimental Biology, Washington, D.C., 1962.
47. Nornes, H. O. and Carry, M., Neurogenesis in spinal cord of mouse: an autoradiographic analysis, *Brain Res.,* 159, 1, 1978.
48. Langman, J. and Haden, C. H., Formation and migration of neuroblasts in the spinal cord of the chick embryo, *J. Comp. Neurol.,* 138, 419, 1970.
49. Fujita, S., Analysis of neuron differentiation in the central nervous system by tritiated thymidine autoradiography, *J. Comp. Neurol.,* 122, 311, 1964.
50. Miale, I. L. and Sidman, R. L., An autoradiographic analysis of histogenesis in the mouse cerebellum, *Exp. Neurol.,* 4, 277, 1961.
51. Haeckel, E., *Anthropogenie ou Histoire de L'Evolution Humaine,* C. Reinwald et Cie, Paris, 1877.

INDEX

A

B

C

G

H

N

Nackengrube of His
   in baboon embryos, 37, 39
   in common marmoset embryos, 81
   in human embryos, 11
Nares, 48, 152
Nasal bones, 153
Nasal pits
   in baboon embryos, 43
   in Chinese hamster embryos, 109—111
   in common marmoset embryos, 81
   in guinea pig embryos, 123, 124
   in pig embryos, 161
   in rabbit embryos, 136
   in Rhesus monkey embryos, 65, 67
   in tree shrew embryos, 169
Nasal placode, 121
Nasal process
   in chick embryos, 179, 180
   in Rhesus monkey embryos, 63
   in sheep embryos, 152
Nasal region, 138, 142
Nasal septum, 183
Nasolacrimal cleft, 63
Nasolacrimal duct, 167
Nasolacrimal groove
   in chick embryos, 181
   in Chinese hamster embryos, 112
   in human embryos, 17
   in pig embryos, 162, 163, 165
   in rabbit embryos, 137
   in Rhesus monkey embryos, 67
   in sheep embryos, 151—153
Nasomaxillary groove, 150
Neck region
   in chick embryos, 184
   in common marmoset embryos, 82
   in human embryos, 29
   in Rhesus monkey embryos, 75
   in tree shrew embryos, 169
Neostriatum, 185
Neural crest
   in chick embryos, 173
   in human embryos, 3, 5, 7
Neural folds
   in baboon embryos, 32, 33
   in Chinese hamster embryos, 104
   in guinea pig embryos, 118
   in human embryos, 3
   in Rhesus monkey embryos, 55, 56
   in sheep embryos, 146
Neural groove, 2, 169
Neural plate
   in chick embryos, 172
   in human embryos, 5
   in pig embryos, 156
   in rabbit embryos, 132
   in sheep embryos, 146
Neural tube
   in chick embryos, 172
   in Chinese hamster embryos, 105
   in guinea pig embryos, 118
   in mouse embryos, 90, 91
   in pig embryos, 157
   in rabbit embryos, 133
   in tree shrew embryos, 169
Neurohypophyseal evagination, 164
Nictitating membrane, 184, 185
Nose, see also Olfactory placode
   in baboon embryos, 47
   in guinea pig embryos, 129
   in human embryos, 19, 23
   in Rhesus monkey embryos, 77
Nostrils, 113, 114
Notochord, 55

O

Olfactory bulb, 164
Olfactory evagination, 17
Olfactory invagination, 176, 177
Olfactory nerves, 181
Olfactory pit
   in baboon embryos, 39
   in mouse embryos, 94, 95
   in pig embryos, 162, 163
   in Rhesus monkey embryos, 61, 63
Olfactory placode
   in baboon embryos, 37, 41
   in chick embryos, 174
   in Chinese hamster embryos, 107, 108, 111
   in common marmoset embryos, 81
   in guinea pig embryos, 120, 122
   in human embryos, 9, 13
   in mouse embryos, 91—93
   in rabbit embryos, 134
   in Rhesus monkey embryos, 59
   in sheep embryos, 147, 149
Olfactory plate, 11
Olfactory region, 82
Opposable digits, see also Digits, 49
Optic cup
   in human embryos, 19
   in pig embryos, 161
   in sheep embryos, 149, 150
Optic nerve, 21, 25, 27, 29
Optic primordia, 2
Optic stalk, 21, 23, 25
Optic vesicles
   in chick embryos, 173, 174
   in common marmoset embryos, 81
   in human embryos, 9
   in pig embryos, 158
   in rabbit embryos, 133
   in sheep embryos, 147
   in tree shrew embryos, 169
Oronasal groove, 127
Oryctolagus cuniculus, see Rabbit embryos
Ossification

S

T